Hallmarks of Parkinson's Disease

Hallmarks of Parkinson's Disease

Editor

Cristine Alves Da Costa

MDPI • Basel • Beijing • Wuhan • Barcelona • Belgrade • Manchester • Tokyo • Cluj • Tianjin

Editor
Cristine Alves Da Costa
Université Côte d'Azur
France

Editorial Office
MDPI
St. Alban-Anlage 66
4052 Basel, Switzerland

This is a reprint of articles from the Special Issue published online in the open access journal *Cells* (ISSN 2073-4409) (available at: https://www.mdpi.com/journal/cells/special_issues/Hallmarksof_Parkinson).

For citation purposes, cite each article independently as indicated on the article page online and as indicated below:

LastName, A.A.; LastName, B.B.; LastName, C.C. Article Title. *Journal Name* **Year**, *Volume Number*, Page Range.

ISBN 978-3-0365-5015-2 (Hbk)
ISBN 978-3-0365-5016-9 (PDF)

Contents

About the Editor

Cristine Alves Da Costa

Cristine Alves Da Costa (PhD) is a research director at INSERM and manages a team at the Institute of Molecular and Cellular Pharmacology (IPMC-CNRS, UMR7275). She guides a team that studies the implications of Parkinson's disease associated proteins in the origin of cell death in neurodegenerative disorders and tumor suppression. Her research interests include Parkinson's and Alzheimer's disease molecular and cell biology, apoptosis, autophagy and ER stress. Moreover, she is also interested in the interplay between these two pathologies via common molecular denominators and the link between neurodegenerative disorders and human brain cancer.

Review

The Endoplasmic Reticulum Stress/Unfolded Protein Response and Their Contributions to Parkinson's Disease Physiopathology

Cristine Alves da Costa *, Wejdane El Manaa, Eric Duplan and Frédéric Checler

INSERM, CNRS, IPMC, Team Labeled "Laboratory of Excellence (LABEX) Distalz", Université Côte d'Azur, 660 Route des Lucioles, Sophia-Antipolis, 06560 Valbonne, France; elmanaa@ipmc.cnrs.fr (W.E.M.); duplan@ipmc.cnrs.fr (E.D.); checler@ipmc.cnrs.fr (F.C.)

* Correspondence: acosta@ipmc.cnrs.fr; Tel.: +33-(49)-3953457

Received: 23 October 2020; Accepted: 16 November 2020; Published: 17 November 2020

Abstract: Parkinson's disease (PD) is a multifactorial age-related movement disorder in which defects of both mitochondria and the endoplasmic reticulum (ER) have been reported. The unfolded protein response (UPR) has emerged as a key cellular dysfunction associated with the etiology of the disease. The UPR involves a coordinated response initiated in the endoplasmic reticulum that grants the correct folding of proteins. This review gives insights on the ER and its functioning; the UPR signaling cascades; and the link between ER stress, UPR activation, and physiopathology of PD. Thus, *post-mortem* studies and data obtained by either *in vitro* and *in vivo* pharmacological approaches or by genetic modulation of PD causative genes are described. Further, we discuss the relevance and impact of the UPR to sporadic and genetic PD pathology.

Keywords: Parkinson's disease; unfolded protein response; reticulum endoplasmic; genetics

---

Parkinson's disease (PD) is the second most frequent neurodegenerative disorder after Alzheimer's disease. It is characterized at a histopathological level by the presence of intracellular lesions named Lewy bodies or Lewy neurites according to their shape and by exacerbated cell death of dopaminergic neurons. The deficit in dopamine caused by the *substantia nigra* neuronal loss is translated clinically by uncontrollable tremor, hypokinesia, spasticity, and gait abnormalities. Multiple pieces of evidence indicate that two cellular organelles are strongly linked to the physiopathology of PD: the mitochondria and the endoplasmic reticulum (ER). This review will discuss the role of the ER and the signaling cascades activated by this organelle during ER stress and how this dysfunction could account for the etiology of PD.

1. The Endoplasmic Reticulum

The endoplasmic reticulum is a cellular organelle that controls the synthesis, the folding, and the post-transductional modifications of almost one-third of proteins. It is the first compartment of the secretion pathway (ER–Golgi–lysosome) in eukaryotic cells. The ER forms a network of elongated tubules and flattened discs covering a large part of the cytoplasm that extends to the nuclear envelope [1,2]. Considering the key role of this organelle in the development of the unfolded protein response (UPR), we provide a short description of its structure and functions below.

1.1. Structure of the ER

The ER is usually categorized as smooth (SER) or rough (RER), depending on its morphology, while its intramembranous space is named lumen. The rough phenotype of the RER is linked to the presence of attached ribosomes at its surface facing cytoplasm. The SER is involved in synthesis of carbohydrates and lipids and the RER in production of membranes and secretory proteins [1,2].

Usually, the SER gathers a tubular network and the RER a series of flattened sacs. More recently, a new classification, which takes into account the structure of the membrane rather than its morphology, has been proposed. According to this new classification, the ER is composed of the nuclear envelope, flattened membrane-enclosed sacs known as cisternae, and an interconnected tubular network [3]. These components of the ER are distinguished by the membrane curve. Thus, the tubules of the ER harbor a more important membrane curvature than that of the leaflets of the nuclear envelope and cisternae. The volume of the ER is cell type-dependent. Nevertheless, the ER occupies a consequent cell volume, allowing it to establish contact sites with several intracellular organelles.

Accordingly, the ER interacts with the mitochondria, the plasma membrane, endosomes, and the endolysosomal system. Thus, the ER associates with the mitochondria via the MAM (mitochondria-associated membrane), allowing the exchange of calcium and lipids between these two key cellular organelles [4]. The ER is also in contact with the plasma membrane via ORA1 (olfactory receptor class A-1 like protein 1), CRACM1 (calcium release-activated calcium channel protein 1) and STIM1 (stromal interaction molecule 1), which are both regulated by calcium and are localized in the plasma membrane and in the ER, respectively [5]. The ER interacts with the endosomes [6] via the sterol binding proteins STARD3 (StAR-related lipid transfer protein 3) and its ER-binding partner STARD3NL (STARD3 N-terminal like protein) [7], allowing the delivery of cholesterol to endosomes. Finally, the ER can also interact with the endolysosomal system via MDM1/SNX13 (mitochondrial distribution and morphology 1/Sorting NeXin 13) [8], suggesting an implication of the ER in autophagy control.

Alone or in coordination with other cell organelles, the ER develops several essential functions that control cellular homeostasis.

1.2. ER Functions

The ER contributes to several physiological functions. Notably, it is involved in the synthesis and storage of lipids; the synthesis, folding, and export of proteins; calcium homeostasis; and the metabolism of glucose [4]. The ER is a dynamic organelle that is sensitive to nutriments and that coordinates energetic fluctuations and the firing of the most adequate metabolic response necessary to maintain the cell homeostasis.

1.2.1. Lipid Synthesis

The ER plays an essential role in membrane synthesis, the synthesis of lipid vesicles, and the accumulation of fat for energy storage. Lipid synthesis takes place at the membrane level, at membrane interfaces, and at ER contact sites with other organelles. The lipid precursors synthesized in the ER membrane are then converted into structural lipids, sterols, steroid hormones, biliary acids, dolichols, prenyl donors, and a myriad of isoprenoid species with key functions for cellular metabolism. The ER dynamically modifies its membrane structure to adapt to variations in cellular lipid concentrations. It also grants cholesterol homeostasis [9] and the synthesis of lipid components of the cell membrane, namely, sterols, sphingolipids, and phospholipids [10].

1.2.2. Export of Proteins and Lipids

Most of the proteins and lipids synthetized in the ER must be transported to other cell structures mainly by the secretory pathway. To maintain a constant normal flux, the export of proteins must be strictly regulated and any failure of the process of secretion may, in return, severely impact the structure and function of the ER. The generation of the ER–Golgi COPII (coat complex II) transport vesicles is at the heart of the lipid export process [11], but other mechanisms have also been described. For example, it has been shown that a great quantity of lipoproteins are exported from the ER via another type of vesicle named prechylomicron [12].

1.2.3. Calcium Homeostasis

The ER is the main storage site and it plays a central role in the regulation of Ca^{2+} intracellular levels. Calcium is toxic for most of the metabolic processes, but it is also a key signal mediator of several cellular processes. The cellular calcium levels should be tightly regulated to allow for the proper development of protein folding and a timely specific release of calcium. Certain regions of the ER are implicated in the fine regulation of calcium concentration, notably the contact zones between the ER–mitochondria (MAMs) and ER–plasma membrane. The ER takes advantage of a coordinated cascade of events to control Ca^{2+} concentration at each side of its membrane. First, a calcium pump present in the ER membrane allows for the entry of Ca^{2+} in the ER; next, chaperone proteins bind and buffer the free Ca^{2+}; and finally, the ER membrane channels grant the release of Ca^{2+} in the cytosol [13].

1.2.4. Synthesis and Folding of Proteins

The main function of the ER is the synthesis and folding of proteins of the secretory pathway, a process mediated by luminal resident chaperones and foldases. These proteins represent 30% of the proteome and are either addressed to the plasma membrane (ionic channels, transporters, etc.), Golgi apparatus, lysosomes (proteases, lipases, etc.), or secreted (albumin, growth factors, insulin, etc.). Some of these proteins may also stay inside the ER as certain chaperones. The folding process includes a translational and post-translational phase in which a newly synthesized protein in the ribosomes (RER) endures a series of modifications and come across a number of molecular chaperones and foldases that assist its proper folding and issue from the ER. The main modifications taking place during the folding process include the cleavage of the signal peptide by the signal sequence peptidase complex (SPC), N-linked glycosylation, formation of disulfide bonds, pro-isomerization, and oligomerization. All modifications taking place during the folding process may be associated with both translational and post translational phases, except oligomerization, which is a post-translational modification. The detailed steps of the folding process have been reviewed by Braakman et al. [14].

The cell consumes a lot of energy to keep the ionic and electronic environment of the ER perfectly adapted to protein folding. Indeed, the ER grants a much higher calcium concentration and a more oxidizing redox potential than cytosol [15]. The resident ER chaperone proteins are the first elements mobilized by the cellular machinery to catalyze the proper folding of the neo-synthetized proteins and they bind and prevent aggregation during the maturation process. These chaperones include the ER lectins (calnexins and calreticulins) and heat shock proteins (HSPs) of the ER (GRP78/BiP (glucose-related protein 78/binding immunoglobulin protein), HSP70 (heat shock protein 70), and GRP94 (glucose-related protein 94)). GRP78/BiP is the most abundant chaperone of the ER [16,17]. Once partially folded, the proteins are taken over by GRP94, which inserts itself into the heart of the protein via an amphipathic finger. GRP94 is a selective chaperone that allows the correct folding of specific proteins. However, the selectivity criteria of GRP94 are still poorly understood.

Often the folding and structural processing of proteins also involves the co-translational addition of an oligosaccharide. This process called *N*-glycosylation is crucial as it ensures that proteins of the secretory pathway are correctly folded, modified, and assembled into multi-protein complexes in the ER. The *N*-glycosylation also prevents the progression of misfolded proteins into the secretory pathway [18]. When the protein has reached a certain degree of folding, the last glucose must be removed by α-glucosidase II. If the protein has not reached its final folding state ("native fold"), it will be taken over by the glucosyltransferase GGT, re-glycosylated, and again fixed by the chaperones calnexin or calreticulin [19,20]. This cycle can be repeated several times before a protein is properly folded or degraded.

Despite the existence of this sophisticated protein folding machinery, the success rate of protein folding is weak for numerous proteins of the secretion pathway. Proteins not properly folded are not tolerated by the cell and are eliminated by two efficient "control/quality" systems. The ERAD (endoplasmic reticulum-associated degradation) is the main degradation pathway of soluble proteins and reticulophagy, which allows the degradation of non-soluble protein aggregates [21]. The ERAD allows for the comeback of unfolded proteins to the cytosol and their consequent ubiquitylation and degradation by the 26S proteasome [22]. The reticulophagy is a selective type of autophagy that allows for the clearance of the ER by the lysosome [23].

Despite the toughness of the folding capacity of the ER, cells often operate near the limits of their secretory capacity. Thus, a wide range of cellular disturbances can affect the efficiency of protein folding in the ER and lead to an accumulation of misfolded proteins within this organelle. This phenomenon is known as ER stress.

2. The ER UPR Response

Alterations in ER functions, such as altered calcium levels, increase oxidative stress, or dysfunction of protein N-glycosylation, causing the accumulation of misfolded proteins in the ER, triggering ER stress. In response to the stress of the ER, signaling pathways grouped under the term UPR (unfolded protein response, Figure 1) are activated to circumscribe this stress.

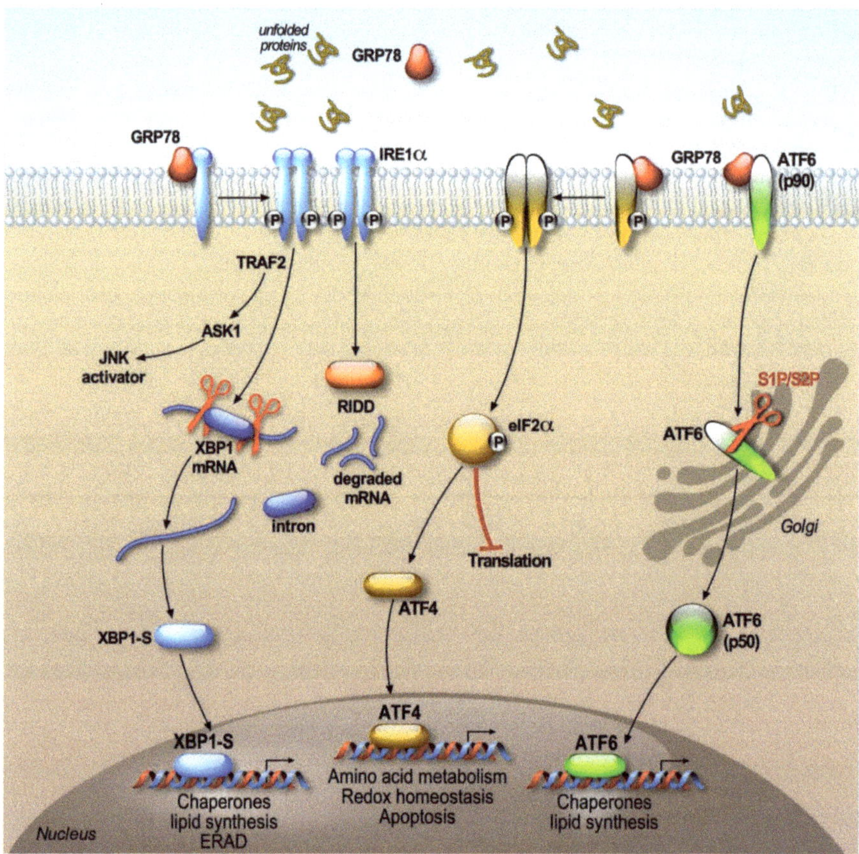

Figure 1. The unfolded protein response (UPR) signaling pathways.

The UPR response concerns an adaptive cellular mechanism that aims to restore the ER homeostasis [24]. The objectives of the activation of the UPR response are (1) to reduce the general synthesis of proteins in order to decrease the accumulation of proteins at the ER lumen, (2) to increase the synthesis of chaperone proteins to facilitate the proper protein folding, and (3) to enhance the translation of proteins implicated in the ERAD machinery in order to foster the elimination of misfolded proteins [25]. The ER stress induces the activation of the three branches of the UPR. The coordinated action of these three branches ultimately boosts the ER folding capacity. Thus, the three branches of the UPR will induce the expression of genes coding either for the chaperones BiP/GRP78, GRP94, and calreticulin, or the enzymes responsible for the establishment of disulfide bridges, in order to improve the ER capacity to properly fold the proteins and thus prevent the accumulation of misfolded proteins [26,27]. In mammals, UPR activation is mediated by signaling pathways involving three sensors located at the ER membrane: IRE1 (inositol-requiring enzyme-1), PERK (protein kinase RNA-like endoplasmic reticulum kinase), and ATF6 (activating transcription factor 6).

The cell activates the IRE1, PERK, and ATF6 pathways, which will differentially modulate downstream effectors capable of binding to specific promoter sequences in order to regulate their target genes. Their luminal domain allows for the integration of the information coming from the lumen of the ER, while their effector cytoplasmic domain mediates the interaction with effectors harboring transcription or translation functions. Thus, the IRE1 effector XBP1 (X box protein 1) and ATF6 bind to sequences ERSE (ER stress response element) [28] in the presence of the cofactor NF-Y (nuclear factor Y), while the transcription factor ATF4 recognizes AARE/CARE (amino acid response element/C/EBP (CAAT-enhancer-binding protein)-ATF) sequences [29]. XBP1 can also recognize UPRE (unfolded protein response) sequences [30].

2.1. BiP, the Trigger of the UPR

In the absence of stress, the IRE1, PERK, and ATF6 sensors are kept inactive due to their association with the chaperone BiP/GRP78. Indeed, the luminal sequences of the three ER sensors interact with BiP, which constitutively prevents the activation of these regulatory pathways [31,32]. When misfolded proteins accumulate in the ER, the BiP/GRP78 protein associates preferentially with malformed proteins and dissociates from IRE1, PERK, and ATF6, thus leading to their activation [31,33]. Thanks to this system, the cell can quickly determine the state of reticular stress as a function of the amount of free or bound BiP [34].

Once detached from BiP, "free" IRE1 and PERK form homodimers or oligomers and self-phosphorylate to activate their downstream targets [31]. However, dissociation of BiP from ATF6 unmasks a motif of export of ER [33] that facilitates ATF6 translocation towards the Golgi apparatus [35]. This model of competition of BiP during the activation of the UPR response indicates that BiP is an inhibitor of the UPR response. However, other BiP-dependent or -independent models have been proposed [36]. In addition, it has recently been reported that another ER lumen chaperone, HSP47, is able to displace BiP from IRE1 to promote its oligomerization [37].

2.2. PERK Signaling

The PERK signaling cascade is the first branch of the UPR mobilized to cope the stress of the ER. It allows for the punctual reduction of protein synthesis. PERK has been identified in the pancreatic islets of rats as a serine/threonine kinase capable of phosphorylating EIF2α (eukaryotic translational initiation factor 2) [26,38]. PERK is an ubiquitously expressed [38] protein that is structurally composed of a luminal and a cytosolic kinase domain [26].

During ER stress, the dissociation of PERK from BiP triggers the oligomerization of PERK [31] and allows its activation by an autophosphorylation process [39]. Activated PERK phosphorylates EIF2α on serine 51 [26]. This phosphorylation inactivates EIF2α and reduces general protein translation. EIF2α is a subunit of the EIF2 heterotrimer that regulates the first step of protein synthesis by promoting the binding of the initiator tRNA (transfer RNA) to the 40S ribosomal subunits [40–42]. Therefore, the blockade of protein translation during ER stress alleviates the folding machinery [43].

Under ER stress conditions, the inactivation of EIF2 leads to the activation of ATF4. ATF4 is a ubiquitous transcription factor that harbors numerous short uORFs (upstream open reading frames) in its 5′ UTR region [43]. The decrease of EIF2 levels upon ER stress leads to a shift of the translation initiation site to the main ORF, resulting in a more efficient synthesis of ATF4 [44]. This process allows increased translation of the transcription factor ATF4, the expression of which is low under physiological conditions [44]. Once synthesized, ATF4 is directed to the nucleus and then binds to the CARE (C/EBP (CAAT-enhancer-binding protein)-ATF-responsive element) of several genes (*ASNS* (asparagine synthetase), *CAT1* (catalase-1), *CHOP* (C/EBP homologous protein), *TRBP3* (transactivation domain binding protein 3), etc.) involved in the control of antioxidant response, protein overload in the ER, and activation of macroautophagy [45].

It is important to note that EIF2α is not the only PERK substrate. Thus, PERK phosphorylates NRF2, allowing its translocation to the nucleus and consequent regulation of genes implicated in anti-oxidant response [46]. The activation of NRF2 by PERK allows the cell to keep its redox potential stable during ER stress.

Finally, PERK is also a mediator of the apoptotic response. In response to acute stress, the PERK pathway modulates several pro-apoptotic factors that contribute to cell degeneration and death if ER stress persists. Thus, even if temporary disruption in protein translation due to phosphorylation of EIF2α is beneficial, sustained blockade of PERK is not compatible with survival of cells [47]. Moreover, the hyperactivation of PERK induces transactivation of CHOP which leads to either a decrease/increase of the expression of anti-apoptotic/pro-apoptotic members of the BCL-2 (B-cell lymphoma 2/BIM (BCL-2-like protein 11) families [47]. The modulation of CHOP by PERK also triggers an increase of oxygenated free radicals by increasing the resident oxidases of the ER such as ERO1A (endoplasmic reticulum oxidoreductase 1 alpha) [48]. CHOP also reduces the total amount of cellular gluthatione and inhibits the translation block linked to PERK-mediated EIF2α phosphorylation [49]. This effect is dependent on GADD34, a phosphatase associated with EIF2α dephosphorylation that is transcriptionally regulated by CHOP [50]. Other pro-apoptotic proteins such as PUMA (p53 upregulated modulator of apoptosis) are also activated by CHOP during acute ER stress [51].

2.3. *ATF6*

ATF6 is an ER transmembrane protein that has a DNA-binding bZIP (basic leucine zip) domain [52]. Under stressful conditions, there is dissociation of the BiP–ATF6 complex, and ATF6 translocates into the Golgi apparatus and is sequentially cleaved by S1P and S2P proteases [53] to produce a cytosolic fragment (ATF6f). This fragment interacts with different nuclear partners, allowing it to upregulate the transcription of ER chaperones and ERAD/EDEM (ER Degradation Enhancing Alpha-Mannosidase-Like) genes [30]. ATF6f collaborates with the cofactor NF-Y (nuclear factor Y) and binds to its targets genes via the recognition of a consensus motif (ER stress response element) [54].

In humans, ATF6 is encoded by two genes: *ATF6A* for ATF6α [54], and *ATF6B* for ATF6β [41]. Despite their high homology, ATF6β behaves as a negative regulator of ATF6α [55]. ATF6's most important targets are BiP, GRP94, and calnexin [28], but ATF6α also interacts with other bZIP proteins, such as CREB (cAMP (Adenosine 3′5′ Cyclic Monophosphate) Response Element Binding Protein), CREB3L3, NF-Y, yin yang 1, and XBP1 [56,57]. ATF6 and XBP1 are known to act synergistically since they can form heterodimers, allowing ATF6α to act as a stimulator of the transcription of members of ERAD harboring UPRE sites recognized by XBP1 [30]. Indeed, EDEMs and ERAD proteins (3-hydroxy-3-methylglutaryl-coenzyme A (HMG-CoA), reductase degradation 1 (HRD1) and Herp (Homocysteine-induced endoplasmic reticulum protein)) are all transactivated by these heterodimers during ER stress [30].

2.4. *IRE1 (Inositol-Requiring Enzyme 1)*

The IRE1 pathway is the most conserved and best studied UPR pathway. IRE1 has two homologs, IRE1α and IRE1β, which share 40% of structural homologies [58–60]. IRE1α is expressed in all cells while the expression of IRE1β is restricted to the gastrointestinal system and to the lining of the pulmonary epithelium [61,62]. The mode of activation of IRE1 is similar to that of PERK. However, besides possessing a kinase activity, the cytosolic domain of IRE1 harbors an atypical endoribonuclease

(RNAse) activity. Like other players of the UPR, IRE1 is constitutively inactive when associated to BiP; however, in response to the accumulation of misfolded proteins in the ER, it dissociates from BiP. IRE1 dissociation from BiP allows its dimerization and autophosphorylation on several serine residues. The exact role of these phosphorylations remains unknown, but three of them (Ser724, Ser726, Ser729) have been described as essential for the complete activation of the endoribonuclease function of IRE1 [63]. Importantly, the endoribonuclase activity of IRE1 is responsible for the unconventional splicing of XBP1 mRNA [64,65]. Thus, the excision of 26 nucleotides (intron) in the mRNA of XBP1 causes a shift in the reading frame during the translation of the mRNA, which introduces a new carboxyl domain in the protein XBP1. The splicing of XBP1 mRNA by IRE1 is considered atypical, since it takes place in the cytoplasm rather than in the nucleus, and does not require the consensus sequences used by the spliceosome [66,67]. This atypical splicing makes it possible to generate a stable and active XBP1 protein known as XBP1-S (XBP1-spliced). XBP1-S is a transcription factor composed of a nuclear compartment targeting signal sequence, a transcriptional activation domain and bZIP DNA-binding and dimerization domains.

Depending on the tissue context and stimuli, XBP1 can interact with other transcription factors, forming heterodimers [68]. Under ER stress conditions, XBP1 controls the expression of factors modulating folding, secretion, ERAD, protein translocation in the ER, and lipid synthesis [69,70].

Besides the alternative splicing of XBP1 mRNA, the RNase domain of IRE1 also regulates the stability of several mRNAs. However, unlike XBP1, they are not spliced to produce mature but degraded mRNAs. This process is known under the name of RIDD (regulated IRE1-dependent decay) [27], and it consists in degrading mRNAs directly localized at the ER membrane that do not contain a signal peptide and a specific secondary structure [71]. Thus, IRE1 can degrade its own mRNA in order to regulate its own activation [58], but also other mRNAs, in order to decrease protein synthesis. Ultimately, the role of IRE1 RIDD activity is to control the translation of proteins requiring complex spatial folding that can potentially burden the ER. Interestingly, the unspliced form of XBP1 (XBPu) encodes a protein that acts as a transcriptional repressor of XBP1 [72]. The unconventional splicing of XBP1 mRNA is regulated at different levels and is linked to the transient expression of the unspliced form of XBP1 (XBP1u).

Although XBP1u is highly unstable and rapidly degraded by the 26S proteasome during translation, it can block ER membrane hooked ribosomes through a well-conserved hydrophobic domain, and as a consequence allows the splicing of XBP1 mRNA in the cytosol [73–75]. Thus, XBP1u sends its own mRNA to the IRE1 splicing site. The selective targeting of XBP1u mRNA to the ER membrane is mediated by a direct interaction of the ER with the Sec61 translocon [76].

The IRE1/XBP1 pathway is mainly a pro-survival signaling pathway. However, the IRE1 pathway can trigger cell death by apoptosis under certain conditions. Indeed, during ER stress, IRE1 recruits the adapter protein TRAF2 (TNF receptor-associated factor 2) to the ER membrane, leading to the activation of ASK1 (apoptosis signal-regulating kinase 1) and its JNK targets (c-Jun NH2 terminal kinase) and p38 MAPK (mitogen-activated protein kinase) [77,78]. Activated JNK can in turn regulate various members of the BCL-2 family, particularly the pro-apoptotic factors BID (BH3 Interacting Domain Death Agonist) and BIM and the anti-apoptotic factors BCL-2, BCL-XL, and MCL-1 (Induced myeloid leukemia cell differentiation protein) [79,80]. Importantly, p38 MAPK phosphorylates and activates CHOP, which increases expression of BIM and DR5 (Death receptor 5), thereby promoting apoptosis [81,82]. Distinct pro-apoptotic proteins such as BAX, BAK, AIP1 (Actin-Interacting Protein 1), and PTP1B (Protein Tyrosine Phosphatase 1 beta) can interact with IRE1 to facilitate its endoribonuclease activity and thus increase the splicing of XBP1 [83–85]. As XBP1 is known as a cytoprotective effector, this regulation suggests that in the early stages of UPR, the possible stimulation of pro-apoptotic factors is not always deleterious and can preserve cellular homeostasis [86,87].

The temporal control of the signaling of the UPR pathway is fundamental in determining the fate of a cell under conditions of ER stress. Although the mechanisms explaining the transition from the adaptive UPR response to the apoptotic UPR response are not definitively established, several models have been proposed to explain how information on the intensity and duration of stimuli is integrated by the cell.

Initially, the UPR response was viewed as a pathway for direct and linear transduction of ER stress levels. However, recent findings have indicated that the three major UPR sensors are finely regulated by post-translational modifications and their binding to various cofactors.

3. Implication of the ER UPR in PD Pathology

The link between the UPR and PD's pathology has been supported by numerous data, which are described below.

3.1. Post-Mortem Evidence

The very first study showing a modulation of UPR mediators in PD human brains was provided by Hoozemans and colleagues [88]. They showed an increase of phospho-PERK and phospho-EIF2α protein levels in the *substantia pars compacta* of human PD samples when compared to age-matched controls. An upregulation of BiP in cingulate gyrus and parietal cortex was also demonstrated in dementia with Lewy bodies (DLB) and PD patients by both Western blot and immunohistochemical approaches. Moreover, the accumulation of pPERK (phospho-PERK) in PD human brain was confirmed by immunohistochemical approaches [89].

More recently, Baek and colleagues showed that the mRNA levels of BiP are increased in several brain regions including the cingulate gyrus. However, in contrast to previous data, they showed a decrease of BiP proteins levels [90]. A modulation of GRP78/BiP, ATF4, and CHOP protein levels was observed in SNpc (*substantia nigra pars compacta*) in *post-mortem* human brain samples [91,92]. The accumulation of PDIp, a member of the protein disulfide isomerase family, in PD human brain tissue corroborates the correlation of UPR activation in PD physiopathology and constitutes a neuroprotective adaptive response against ER stress [93]. The PDI family of proteins are linked to disulfide bond formation, reduction, or isomerization of nascent proteins [94,95]. They grant the accurate folding of proteins and are activated during the UPR [96]. Importantly this protein is nitrosylated in PD, leading to its loss of function [97]. An increase of the levels of phosphorylated IRE1 in PD patient samples indicated that the IRE1-XBP1 signaling is associated with PD pathology [98]. Finally, it has been shown that the levels of HERP, a stress response protein associated with ER folding, ER load reduction, and ERAD-mediated degradation of proteins was found to be increased in the *substantia nigra* of PD individuals [99].

3.2. Pharmacological Approaches In Vitro and In Vivo

The first evidence of a cause–effect link between ER stress and PD was obtained by pharmacological modulation of UPR *in vitro*. Thus, several studies have shown that treatment of different cellular models, notably the dopaminergic SH-SY5Y neuroblastoma cell line, leads to increased ER stress response. Thus, the parkinsonian inducers 6-hydroxydopamine (6OHDA), 1-methyl-4-phenyl-pyridinium (MPP^+), and rotenone trigger a significant increase in transcripts associated with the unfolded protein response [100–102] in various cell models. This transcriptional regulation was corroborated by the post-transcriptional modulation of the key ER stress kinases IRE1α and PERK and their downstream targets [100]. Microarray analysis of MN9D cells treated with 6OHDA and MPP^+ confirmed the regulation of transcripts linked to the UPR and showed that both drugs induced a huge upregulation of the pro-apoptotic-linked transcription factor CHOP [101]. Calcium alterations, BiP (decrease), and CHOP (increase) protein level modulation were evidenced in SH-SY5Y cells treated with MPP^+ [103].

Interestingly, MPP^+ was shown to induce CDK5 (Cyclin-dependent-like kinase 5)-mediated phosphorylation of XBP1s in rat primary cultured neurons. This phosphorylation favored its nuclear shuttle and transcriptional activity, reinforcing the role of the IRE1-XBP1 pathway in the pathogenesis of sporadic PD [104].

Upregulated levels of phosphorylated EIF2α, BiP, and CHOP expression was evidenced in human and rat dopaminergic models submitted to a 6OHDA treatment [105–107]. Moreover, like rotenone, paraquat, 6OHDA, another toxin linked to sporadic PD, was shown to induce apoptosis via the activation of the IRE1α branch of UPR in human and mouse dopaminergic cells [108,109].

Importantly, several animal studies corroborate the *in vitro* data described above. Thus, an induction of the pro-apoptotic IRE1α/caspase-12 branch of the UPR has been shown in the rotenone rat model of PD [110,111]. The systemic delivery of MPTP (1-methyl-4-phenyl-1,2,3,6-tetrahydropyridine) to mice triggers an induction of BiP and CHOP protein and mRNA levels [92], while the intracerebral injection of its metabolite MPP^+ in rabbit brain leads an activation of ATF6 pathway in SNpc [112]. An induction of the proteins levels of GRP78, CHOP and caspase-12 was reproduced in the model of 6OHDA lesion in rats [113].

Interestingly, the injection of the ER inducer tunicamycin *in vivo* into mice brains caused locomotor deficiency, the death of dopaminergic neurons, and activation of the glia [114]. In addition, high levels of oligomerized α-synuclein was observed in the SNpc of animals injected with tunicamycin. These results suggest that administration of tunicamycin into the *substantia nigra* could be a particularly relevant new pharmacological model of PD for examining the impact of ER stress *in vivo*.

3.3. PD Gene Products and Their Influence on the UPR

Genes responsible for autosomal dominant forms of PD.

Of utmost importance, a molecular correlation between PD and the UPR came from studies implying several autosomal-dominant (AD) and -recessive (AR) PD causative genes. Most studies regarding the implication of AD–PD causative proteins in UPR regulation are linked to α-synuclein.

α-Synuclein is a protein encoded by the *SNCA* gene that accumulates in Lewy bodies and Lewy neurites. Several point mutations, duplications, and triplication of the gene have been identified, and multiple *in vitro* and *in vivo* studies indicate that its accumulation triggers its aggregation and thereby induces neurotoxicity [115,116]. The accumulation of aggregated α-syn in the brain and notably its soluble oligomeric toxic form is strongly linked to the etiology of PD [117,118].

The overexpression of α-syn, and thus its aggregated toxic forms, correlates with the chronic activation of multiple branches of the UPR and ER stress-mediated apoptosis. Thus, it has been shown that the overexpression of α-syn triggers the activation of UPR in yeast [119], and that its phosphorylation at serine 129, which is associated to its aggregation and toxicity, leads to an important ATF6 regulation in dopaminergic *in vitro* models [120]. Wild-type and mutated α-syn overexpression in SH-SY5Y cells triggers an alteration in calcium metabolism and an activation of IRE1α-XBP1-signaling pathway [121]. The treatment of differentiated SH-SY5Y cells with oligomeric but not monomeric α-syn leads to enhanced XBP1 splicing, indicating a specific activation of the IRE1-XBP1 signaling pathway by α-syn oligomers [122]. Differentiated 3D5 human neuroblastoma-derived cells overexpressing α-syn show increased levels of GRP78/BiP and phospho-EIF2α [123] in basal conditions, and tunicamycin-induced ER stress leads to accumulation of oligomeric α-syn [123], indicating that ER stress may feed α-syn aggregation and toxicity. α-Syn crowding within the ER induces the activation of the PERK-dependent pathway of the UPR *in vitro* and *in vivo*, an activation process mediated by α-synuclein direct interaction with BiP UPR [124]. α-Synuclein affects ATF6 processing directly via protein–protein interactions or indirectly by means of the reduced incorporation to COPII vesicles. Altered ATF6 processing leads to an impairment of ERAD and increased apoptotic response [125].

Mutations of α-syn that trigger α-syn aggregation affect the UPR response. Thus, the A30P α-syn mutation impacts the mRNA levels of genes involved in the UPR *in vitro* and *in vivo* and induces Golgi fragmentation in LUHMES (Lund Human Mesencephalic) cells [126], while the overexpression of A53T α-syn mutation upregulates the levels of BiP and phosho-EIF2α [127].

The modulation of the UPR by α-syn is not restricted to neurons since mutated α-syn was shown to activate the PERK axis in astrocytes [128]. Considering that astrocytes are involved in various brain functions and support neuronal activity, an activation of UPR by α-syn in these cells may lead to deleterious consequences.

This network of evidence does not make it possible to ascertain whether α-syn neurotoxicity is the cause or the consequence of UPR failure and thus which of them is the primary trigger of PD pathogenesis. Nevertheless, a recent work from Colla et al. in A53T α-syn transgenic mice indicates that the accumulation of α-syn toxic species in the ER is responsible for UPR activation [129] and that the detection of ER-associated α-syn oligomers precedes ER stress response [130], thus suggesting that UPR activation is rather the consequence of accumulation of α-syn in PD. The aggregation of α-syn in the ER has been corroborated by approaches implying (fluorescence resonance energy transfer) FRET biosensors [131].

LRRK2 (leucine-rich repeat kinase 2 gene) is a kinase of the ROCO family [132], the mutations on which are associated to autosomal dominant forms of PD and more than 3% of sporadic PD forms [133–135]. The biological functions of LRRK2 remain poorly understood, but a few studies suggest that it is linked to UPR. Thus, studies on LRRK2 subcellular distribution in control versus idiopathic PD revealed that LRRK2 is mainly detected in the ER of neurons and that it co-localizes with two ER-specific markers [136]. The analysis of the contribution of a short hairpin RNA (shRNA)-mediated LRRK2 depletion in SH-SY5Y cells leads to a downregulation of BiP in 6OHDA ER stress conditions, indicating that LRRK2 depletion promotes cytoprotection by modulating the UPR [137].

Moreover, recently it has been demonstrated that LRRK2 may affect mitochondrial bioenergetics by modulating ER–mitochondria tethering via the PERK-mediated ubiquitination pathway [138] and that mutated LRRK2-increased ER stress and apoptosis by disabling the sarco/endoplasmic reticulum Ca^{2+}-ATPase (SERCA) in astrocytes [139]. LRRK2-mediated SERCA dysfunction leads to Ca^{2+} overload in the mitochondria.

3.3.1. Genes Responsible for Autosomal Recessive forms of PD (AR PD)

Most studies linking AR PD genes to ER function and UPR are centered around parkin (PRKN), PINK1 (PTEN (Phosphatase and tensin homolog) -induced putative kinase 1) and DJ-1.

Parkin is an E3-ligase [140,141] and transcription factor [142] involved in multiple cellular processes that are affected in PD. Parkin protein is encoded by the PRKN gene, the mutations of which are responsible for most of autosomal recessive juvenile PD [140]. One of the first pieces of evidence linking parkin to ER stress came from *in vitro* studies showing that parkin induced neuroprotection against ER stress [141] and that the Pael (parkin-associated endothelin receptor-like) receptor that is involved in ER stress-mediated apoptosis is a parkin substrate [143]. These studies also demonstrate that the potent ER stress inducer, tunicamycin, leads to an upregulation of parkin mRNA and protein levels that correlates to increased neuroprotection [141]. Moreover, parkin overexpression was found to be protective against Pael ER stress-mediated apoptosis [143].

Interestingly, it has been shown that parkin expression may be induced by either ER or mitochondrial stress via its transcriptional regulation by ATF4. An upregulation of parkin levels protects against mitochondrial failure and cell death, suggesting a functional link between parkin, ER stress, and mitochondrial homeostasis [144]. Moreover, it was shown that salubrinal, an ER stress inhibitor, prevents rotenone-induced apoptosis in SH-SY5Y, corroborating the neuroprotective role of the ATF4–parkin pathway in ER stress triggered by PD inducers [145]. Corroborating the link between ER and mitochondria via parkin, researchers showed that the increase of parkin levels facilitated the crosstalk between these organelles and granted the calcium mitochondrial load to assure cell bioenergetics [146].

We have shown that endogenous and overexpressed parkin are induced by ER stress and that parkin impacts the UPR response via a p53-dependent transcriptional control of XBP1 [147]. These data provide a direct evidence of a role of parkin in neuronal control of the UPR. Of note, parkin-mediated control of ER stress is not restricted to neurons since astrocytes depleted in parkin show increased levels of spliced XBP1, ATF6, ATF4, CHOP, and Ccl2 in response to thapsigargin [148]. Interestingly, it has been shown that the induction of parkin levels may vary according to the cell type since an increased expression of parkin was observed in astrocytes and not primary hippocampal neurons submitted to ER stress [149]. The contrasting data between SH-SY5Y cells and hippocampal neurons may suggest a preponderant function of parkin in dopaminergic neurons. Moreover, in corroborating a cell type-specific induction of parkin by ER stress, it was shown that 2-mercaptoethanol and tunicamycin increased the expression of parkin in SH-SY5Y (H) cells, Neuro2a cells, Goto-P3 cells, but not in SH-SY5Y (J) cells and IMR32 cells [150].

Several *in vivo* models corroborate the impact of parkin to UPR control. Thus, parkin mutant flies show an activation of the PERK branch of the UPR through the establishment of mitofusin bridges between defective mitochondria and the ER [151]. Moreover, drosophila models of parkin overexpression show an enhancement of K48-linked polyubiquitin and reduced levels of protein aggregation during aging [152].

A few studies have implicated PINK1 in the UPR response. PINK1 is a mitochondrial serine/threonine kinase that, in conjunction with parkin, is strongly implicated in the control of mitophagy [153,154]. Mutations of PINK1 are associated to both genetic and sporadic PD cases [155,156] and perturbed mitochondrial homeostasis. Further, the overexpression of the deletion mutant of OTC (ornithine transcarbamylase) (ΔOTC), which induces mitochondrial UPR in mammalian cells [157], leads to an increase of PINK1 protein levels, parkin recruitment, and mitophagy firing without dissipation of mitochondrial potential in HeLa cells [158]. These data indicate that mitochondrial UPR leads to the induction of PINK1–parkin-dependent mitophagy followed by reduced misfolded protein load. Interestingly, PINK1 modulation was also shown to regulate mitochondrial UPR. Thus, mutations in both human and fly PINK1 result in higher levels of misfolded components of respiratory complexes and accumulation of HSP60 [159].

PINK1 was shown to prevent ER-induced apoptosis in mice primary cortical neurons [160], and transcriptomic studies performed in PINK1 knockout aged mice indicated a downregulation of ER stress response genes [161]. Finally, *in vivo* studies in *Drosophila* show that PINK1 mutations are associated with PERK modulation [151].

DJ-1 (PARK7) is a multifunctional protein [162] considered as a mitochondrial oxidative stress cellular sensor that interestingly harbors chaperone properties [163]. In addition to its key mitochondrial function, downregulation of DJ-1 was shown to affect ER mitochondria contacts in SH-SY5Y differentiated cells [164]. Corroborating these data, DJ-1 overexpression was shown to overcome the p53-induced mitochondrial calcium uptake failure and the perturbations in ER–mitochondria tethering [165]. Overexpressed and endogenous DJ-1 proteins protect against ER stress induced by thapsigargin and tunicamycin in Neuro 2a cells [166].

DJ-1 regulates and is regulated by UPR signaling pathway members. Thus, DJ-1 regulates the UPR and apoptotic response through the increase of ATF4 signaling in stress conditions [167] and is transcriptionally regulated by XBP1. Thus, we have shown that XBP-1 directly binds to its promoter, leading to its upregulation [147]. Finally, it has been shown that oxidized DJ-1 binds to R-HSP5 and favors the elimination of misfolded cargo proteins by autophagy in oxidative stress conditions [168].

Among genetic PD, PARK20 is a rare autosomal recessive juvenile Parkinson's form due to mutations in the phosphatidylinositol phosphatase, synaptojanin1 (Synj1) [169,170]. PARK20 fibroblasts show alterations in the exit machinery of the ER and Golgi trafficking. These alterations lead to the activation of the PERK branch of UPR due to the accumulation of cargo proteins in the ER [171].

Finally, mutations in PLA2G6 (calcium-independent phospholipase A2), which are linked to PARK14-linked young-onset dystonia-parkinsonism syndrome with recessive inheritance [172] were shown to upregulate GRP78, IRE1, PERK, and CHOP protein levels *in vivo* [173].

3.3.2. PD Risk Factors

Glucocerebrosidase (GCase, GBA) is a lysosomal enzyme encoded by the GBA gene that is considered an important risk factor to PD [174]. Mutations in GCase are associated to α-syn accumulation due to an impairment of its CMA (chaperone-mediated autophagy)-mediated degradation [175]. *Post-mortem* analysis of brains of Lewy bodies dementia (LDB) patients carrying GBA1 mutations show alterations on protein levels BiP and HERP, indicating abnormal UPR response [176]. Horowitz's team has shown that mutations of GCase lead to their retention in the ER and subsequent activation of the UPR in the *Drosophila* model [177]. Moreover, they showed that the activation of UPR, illustrated by increased mRNA levels of XBP1s and Hsp-70, may be reversed by ambraxol, a GCase chaperone [178].

Even if it is still debated, high-temperature requirement A2 (HTRA2/Omi/PARK13) is often considered as a PD risk factor [179,180]. HTRA2 is a serine protease with strong homology to the *Escherichia coli* HTRA2, that are important to bacterial survival at high temperatures. Considering that bacterial HTRA2 is involved in the elimination of misfolded aggregated proteins, it is not surprising that HTRA2 is functionally linked to the UPR. Thus, it has been shown that HTRA2 depletion/invalidation in SH-SY5Y and immortalized mouse embryonic fibroblasts (MEFs) triggers a decrease of the pro-apoptotic CHOP protein in 6OHDA stress conditions [181,182]. Interestingly, it has been shown that HTRA2 is induced by tunicamycin *in vitro*, indicating that Omi is activated by ER stress [183].

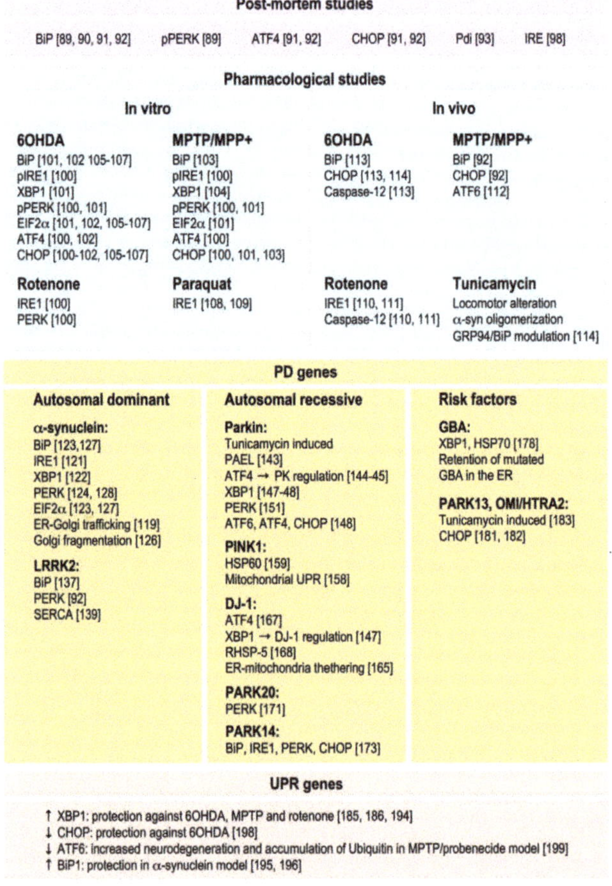

Figure 2. Evidence of the implication of UPR in Parkinson's disease (PD) physiopathology demonstrated by *post-mortem* analysis and *in vitro* and *in vivo* pharmacological/genetic studies. Reference numbers are provided in brackets.

3.4. UPR Gene Products and Their Contribution to PD

Several studies have demonstrated the impact direct of UPR key players to PD physiopathology. Thus, the overexpression of ATF4 by rAAV (Recombinant adeno-associated virus) approaches in a human α-syn rat model of neurodegeneration triggered a severe nigrostriatal cell death due to an activation of caspases 3 and 7 [184].The depletion of XBP1 by shRNA approach in the *substantia nigra* of adult mice triggers chronic stress of the ER and the specific degeneration of dopaminergic neurons. Conversely, rescue of XBP1 level by gene therapy increases neuronal survival and reduces striatal denervation induced by 6OHDA treatment [185]. This study showed the crucial role of the transcription factor XBP1 in controlling the survival of dopaminergic neurons and the vulnerability of dopaminergic neurons to misfolded proteins. Similar results were obtained in mice after administration of MPTP, or in neuroblastoma cell lines treated with MPTP or proteasome inhibitors [186]. In both cases, the overexpression of XBP1 protects the dopaminergic neurons. Interestingly, several studies have shown that these adaptative responses can be stimulated by preconditioning treatments that confer resistance to a subsequent toxic challenge, a phenomenon called "hormesis" [187,188]. Thus, Mollereau and colleagues demonstrated that the preconditioning of the ER leads to neuroprotection in animal models of PD [189,190]. Interestingly, it has been shown that in the XBP1 conditional knockout animal model, XBP1 depletion pre-conditions dopaminergic neurons to stress, rendering them more resistant to 6OHDA treatment. This protection is accompanied by an increase in the expression of markers of the adaptive response of UPR in SNpc. This preconditioning effect is similar to that demonstrated in mice and *Drosophila* by pharmacological approaches where low doses of tunicamycin selectively induce an adaptive UPR response, involving the expression of XBP1-S and not the apoptotic factor CHOP, offering protection of dopaminergic neurons against 6OHDA challenge [191].

Atypical XBP1 splicing is catalyzed by endoribonuclease IRE1 and RTCB1 (RNA 2′,3′-cyclic phosphate and 5′-OH ligase)-ligase [192]. This ligase has been shown to protect dopaminergic neurons from the effects of overexpression of α-syn in *Caenorhabditis elegans*. This observation made it possible to discover a functional relationship between XBP1 and this ligase in the regulation of neuroprotection against proteostatic stress in these neurons [193]. Furthermore, XBP1 has been shown to be not only be protective when delivered to dopaminergic cells by viral transduction but also when transfected into neural stem cells [194]. In these cells, transfection of XBP1 leads to increased survival and improved motor deficits in rat models of PD, injected with rotenone [194]. One of the functions of XBP1 is to associate with ATF6f to enable transcription of the BiP chaperones. Overexpression of this chaperone also protects dopaminergic neurons and improves motor performance in rat models of PD, induced by direct injection of (adeno-associated virus) AAVs encoding the human form of α-syn into SNpc [195]. This protection is accompanied by an overall reduction in the stress response of ER [195]. Age-related decline in BiP or siRNA expression against BiP has also been shown to increase the vulnerability of neurons to α-syn in the same model of PD [196].

3-Hydroxy-3-methylglutaryl-coenzyme A (HMG-CoA) reductase degradation 1 (HRD1), a key player of the ERAD machinery, inhibits cell death induced by 6OHDA in SH-SY5Y cells [197]. Furthermore, CHOP invalidation in mice leads to the protection of dopaminergic neurons against 6OHDA [198], and ATF6 depletion fosters neurodegeneration and ubiquitin accumulation upon chronic injection of mice with MPTP/probenecide [199].

Overall, the studies described above indicate that the genetic modulation of UPR players may lead to novel therapeutic strategies based on the development of pharmacological modulators of gene products of the UPR.

4. Concluding Remarks

The numerous studies described above and resumed in Figure 2 highlight the importance of the ER UPR in the physiopathology of PD. They indicate that all branches of the UPR are likely implicated in PD etiology, but the exact chronology of their activation and hierarchy of their pathogenic weights in human brain remain to be established. It is worth noting that the studies implying PD gene modulation in cellular and animal models have strongly contributed to the delineation of UPR signaling cascades underlying neurodegeneration in PD and have reinforced the functional link between the ER and mitochondria. Several studies have highlighted the importance of mitochondrial UPR and the MAMs in this cellular crosstalk. The implication of mitochondrial UPR in PD has been recently reviewed [200,201]. Interestingly, the main pieces of evidence linking mitochondrial UPR to PD pathology came from functional studies linked to PD-related proteins. Notably, it has been shown that misfolded α-synuclein accumulates in the mitochondria [202,203]; that PINK1 interacts with TRAP2, HTRA2, and HSP60 [204–206]; and that HTRA2 depletion leads to increased levels of CHOP [182]. It also remains unclear as to whether the UPR dysfunction is rather a cause or consequence of PD; however, there is a general consensus that short and mild UPR activation is beneficial while its sustained activation would be deleterious.

Finally, as a corollary of these fundamental studies that put the UPR at the frontline of cellular dysfunctions taking place in PD, many applied/therapeutic works have recently emerged and are reviewed in [207–209]. These works indicate that the development of either pharmacological or genetic strategies to increase the buffering capacity of the proteostasis network may be clinically relevant at short- to mid-term levels. Future fundamental studies should contribute to a better understanding of the UPR mechanism dysfunctions in PD and allow for the development of new therapeutic approaches.

Funding: This research received no external funding.

Acknowledgments: We would like to thank the France-Parkinson foundation for W.E.M.'s financial support, as well as F. Aguila for the artwork.

Conflicts of Interest: The authors declare no conflict of interest.

References

1. Baumann, O.; Walz, B. Endoplasmic reticulum of animal cells and its organization into structural and functional domains. *Int. Rev. Cytol.* **2001**, *205*, 149–214.
2. Voeltz, G.K.; Rolls, M.M.; Rapoport, T.A. Structural organization of the endoplasmic reticulum. *EMBO Rep.* **2002**, *3*, 944–950. [CrossRef]
3. Shibata, Y.; Voeltz, G.K.; Rapoport, T.A. Rough sheets and smooth tubules. *Cell* **2006**, *126*, 435–439. [CrossRef]
4. Alberts, B.J.A.; Lewis, J.; Raff, M.; Roberts, K.; Walter, P. *Molecular Biology of the Cell*, 4th ed.; Garland Science: New York, NY, USA, 2002.
5. Toulmay, A.; Prinz, W.A. Lipid transfer and signaling at organelle contact sites: The tip of the iceberg. *Curr. Opin. Cell Biol.* **2011**, *23*, 458–463. [CrossRef]
6. Rowland, A.A.; Chitwood, P.J.; Phillips, M.J.; Voeltz, G.K. ER contact sites define the position and timing of endosome fission. *Cell* **2014**, *159*, 1027–1041. [CrossRef] [PubMed]
7. Wilhelm, L.P.; Tomasetto, C.; Alpy, F. Touche! STARD3 and STARD3NL tether the ER to endosomes. *Biochem. Soc. Trans.* **2016**, *44*, 493–498. [CrossRef] [PubMed]
8. Henne, W.M.; Zhu, L.; Balogi, Z.; Stefan, C.; Pleiss, J.A.; Emr, S.D. Mdm1/Snx13 is a novel ER-endolysosomal interorganelle tethering protein. *J. Cell Biol.* **2015**, *210*, 541–551. [CrossRef] [PubMed]
9. Brown, M.S.; Goldstein, J.L. A proteolytic pathway that controls the cholesterol content of membranes, cells, and blood. *Proc. Natl. Acad. Sci. USA* **1999**, *96*, 11041–11048. [CrossRef] [PubMed]
10. Patwardhan, G.A.; Beverly, L.J.; Siskind, L.J. Sphingolipids and mitochondrial apoptosis. *J. Bioenerg. Biomembr.* **2016**, *48*, 153–168. [CrossRef]
11. Barlowe, C.; Orci, L.; Yeung, T.; Hosobuchi, M.; Hamamoto, S.; Salama, N.; Rexach, M.F.; Ravazzola, M.; Amherdt, M.; Schekman, R. COPII: A membrane coat formed by Sec proteins that drive vesicle budding from the endoplasmic reticulum. *Cell* **1994**, *77*, 895–907. [CrossRef]
12. Siddiqi, S.; Saleem, U.; Abumrad, N.A.; Davidson, N.O.; Storch, J.; Siddiqi, S.A.; Mansbach, C.M., 2nd. A novel multiprotein complex is required to generate the prechylomicron transport vesicle from intestinal ER. *J. Lipid Res.* **2010**, *51*, 1918–1928. [CrossRef] [PubMed]
13. Berridge, M.-J. The Endoplasmic Reticulum: A Multifonctional signaling organelle. *Cell Calcium* **2002**, *32*, 235–249. [CrossRef] [PubMed]

14. Braakman, I.; Hebert, D.N. Protein folding in the endoplasmic reticulum. *Cold Spring Harb. Perspect. Biol.* **2013**, *5*, a013201. [CrossRef] [PubMed]
15. Van Anken, E.; Braakman, I. Versatility of the endoplasmic reticulum protein folding factory. *Crit. Rev. Biochem. Mol. Biol.* **2005**, *40*, 191–228. [CrossRef]
16. Gething, M.J. Role and regulation of the ER chaperone BiP. *Semin. Cell Dev. Biol.* **1999**, *10*, 465–472. [CrossRef]
17. Melnick, J.; Dul, J.L.; Argon, Y. Sequential interaction of the chaperones BiP and GRP94 with immunoglobulin chains in the endoplasmic reticulum. *Nature* **1994**, *370*, 373–375. [CrossRef]
18. Wormald, M.R.; Dwek, R.A. Glycoproteins: Glycan presentation and protein-fold stability. *Structure* **1999**, *7*, R155–R160. [CrossRef]
19. Sousa, M.C.; Ferrero-Garcia, M.A.; Parodi, A.J. Recognition of the oligosaccharide and protein moieties of glycoproteins by the UDP-Glc:glycoprotein glucosyltransferase. *Biochemistry* **1992**, *31*, 97–105. [CrossRef]
20. Taylor, S.C.; Ferguson, A.D.; Bergeron, J.J.; Thomas, D.Y. The ER protein folding sensor UDP-glucose glycoprotein-glucosyltransferase modifies substrates distant to local changes in glycoprotein conformation. *Nat. Struct. Mol. Biol.* **2004**, *11*, 128–134. [CrossRef]
21. Bernales, S.; McDonald, K.L.; Walter, P. Autophagy counterbalances endoplasmic reticulum expansion during the unfolded protein response. *PLoS Biol.* **2006**, *4*, e423. [CrossRef]
22. Smith, M.H.; Ploegh, H.L.; Weissman, J.S. Road to ruin: Targeting proteins for degradation in the endoplasmic reticulum. *Science* **2011**, *334*, 1086–1090. [CrossRef] [PubMed]
23. Klionsky, D.J.; Abdelmohsen, K.; Abe, A.; Abedin, M.J.; Abeliovich, H.; Acevedo Arozena, A.; Adachi, H.; Adams, C.M.; Adams, P.D.; Adeli, K.; et al. Guidelines for the use and interpretation of assays for monitoring autophagy (3rd edition). *Autophagy* **2016**, *12*, 1–222. [CrossRef] [PubMed]
24. Hetz, C.; Martinon, F.; Rodriguez, D.; Glimcher, L.H. The unfolded protein response: Integrating stress signals through the stress sensor IRE1alpha. *Physiol. Rev.* **2011**, *91*, 1219–1243. [CrossRef] [PubMed]
25. Schroder, M.; Kaufman, R.J. ER stress and the unfolded protein response. *Mutat. Res.* **2005**, *569*, 29–63. [CrossRef] [PubMed]
26. Harding, H.P.; Zhang, Y.; Ron, D. Protein translation and folding are coupled by an endoplasmic-reticulum-resident kinase. *Nature* **1999**, *397*, 271–274. [CrossRef] [PubMed]
27. Hollien, J.; Weissman, J.S. Decay of endoplasmic reticulum-localized mRNAs during the unfolded protein response. *Science* **2006**, *313*, 104–107. [CrossRef]
28. Yoshida, H.; Haze, K.; Yanagi, H.; Yura, T.; Mori, K. Identification of the cis-acting endoplasmic reticulum stress response element responsible for transcriptional induction of mammalian glucose-regulated proteins. Involvement of basic leucine zipper transcription factors. *J. Biol. Chem.* **1998**, *273*, 33741–33749. [CrossRef]
29. Okada, T.; Yoshida, H.; Akazawa, R.; Negishi, M.; Mori, K. Distinct roles of activating transcription factor 6 (ATF6) and double-stranded RNA-activated protein kinase-like endoplasmic reticulum kinase (PERK) in transcription during the mammalian unfolded protein response. *Biochem. J.* **2002**, *366*, 585–594. [CrossRef]
30. Yamamoto, K.; Sato, T.; Matsui, T.; Sato, M.; Okada, T.; Yoshida, H.; Harada, A.; Mori, K. Transcriptional induction of mammalian ER quality control proteins is mediated by single or combined action of ATF6alpha and XBP1. *Dev. Cell* **2007**, *13*, 365–376. [CrossRef]
31. Bertolotti, A.; Zhang, Y.; Hendershot, L.M.; Harding, H.P.; Ron, D. Dynamic interaction of BiP and ER stress transducers in the unfolded-protein response. *Nat. Cell Biol.* **2000**, *2*, 326–332. [CrossRef]
32. Morris, J.A.; Dorner, A.J.; Edwards, C.A.; Hendershot, L.M.; Kaufman, R.J. Immunoglobulin binding protein (BiP) function is required to protect cells from endoplasmic reticulum stress but is not required for the secretion of selective proteins. *J. Biol. Chem.* **1997**, *272*, 4327–4334. [CrossRef] [PubMed]
33. Shen, J.; Chen, X.; Hendershot, L.; Prywes, R. ER stress regulation of ATF6 localization by dissociation of BiP/GRP78 binding and unmasking of Golgi localization signals. *Dev. Cell* **2002**, *3*, 99–111. [CrossRef]
34. DuRose, J.B.; Tam, A.B.; Niwa, M. Intrinsic capacities of molecular sensors of the unfolded protein response to sense alternate forms of endoplasmic reticulum stress. *Mol. Biol. Cell* **2006**, *17*, 3095–3107. [CrossRef] [PubMed]
35. Ye, J.; Rawson, R.B.; Komuro, R.; Chen, X.; Dave, U.P.; Prywes, R.; Brown, M.S.; Goldstein, J.L. ER stress induces cleavage of membrane-bound ATF6 by the same proteases that process SREBPs. *Mol. Cell* **2000**, *6*, 1355–1364.
36. Carrara, M.; Prischi, F.; Ali, M.M. UPR Signal Activation by Luminal Sensor Domains. *Int. J. Mol. Sci.* **2013**, *14*, 6454–6466. [CrossRef]

37. Sepulveda, D.; Rojas-Rivera, D.; Rodriguez, D.A.; Groenendyk, J.; Kohler, A.; Lebeaupin, C.; Ito, S.; Urra, H.; Carreras-Sureda, A.; Hazari, Y.; et al. Interactome Screening Identifies the ER Luminal Chaperone Hsp47 as a Regulator of the Unfolded Protein Response Transducer IRE1alpha. *Mol. Cell* **2018**, *69*, 238–252.e237. [CrossRef]
38. Shi, Y.; Vattem, K.M.; Sood, R.; An, J.; Liang, J.; Stramm, L.; Wek, R.C. Identification and characterization of pancreatic eukaryotic initiation factor 2 alpha-subunit kinase, PEK, involved in translational control. *Mol. Cell Biol.* **1998**, *18*, 7499–7509. [CrossRef]
39. McQuiston, A.; Diehl, J.A. Recent insights into PERK-dependent signaling from the stressed endoplasmic reticulum. *F1000Research* **2017**, *6*, 1897. [CrossRef]
40. Lloyd, M.A.; Osborne, J.C., Jr.; Safer, B.; Powell, G.M.; Merrick, W.C. Characteristics of eukaryotic initiation factor 2 and its subunits. *J. Biol. Chem.* **1980**, *255*, 1189–1193.
41. Ernst, H.; Duncan, R.F.; Hershey, J.W. Cloning and sequencing of complementary DNAs encoding the alpha-subunit of translational initiation factor eIF-2. Characterization of the protein and its messenger RNA. *J. Biol. Chem.* **1987**, *262*, 1206–1212.
42. Adams, S.L.; Safer, B.; Anderson, W.F.; Merrick, W.C. Eukaryotic initiation complex formation. Evidence for two distinct pathways. *J. Biol. Chem.* **1975**, *250*, 9083–9089. [PubMed]
43. Harding, H.P.; Zhang, Y.; Bertolotti, A.; Zeng, H.; Ron, D. Perk is essential for translational regulation and cell survival during the unfolded protein response. *Mol. Cell* **2000**, *5*, 897–904. [CrossRef]
44. Vattem, K.M.; Wek, R.C. Reinitiation involving upstream ORFs regulates ATF4 mRNA translation in mammalian cells. *Proc. Natl. Acad. Sci. USA* **2004**, *101*, 11269–11274. [CrossRef] [PubMed]
45. Ye, J.; Kumanova, M.; Hart, L.S.; Sloane, K.; Zhang, H.; De Panis, D.N.; Bobrovnikova-Marjon, E.; Diehl, J.A.; Ron, D.; Koumenis, C. The GCN2-ATF4 pathway is critical for tumour cell survival and proliferation in response to nutrient deprivation. *EMBO J.* **2010**, *29*, 2082–2096. [CrossRef]
46. Cullinan, S.B.; Zhang, D.; Hannink, M.; Arvisais, E.; Kaufman, R.J.; Diehl, J.A. Nrf2 is a direct PERK substrate and effector of PERK-dependent cell survival. *Mol. Cell Biol.* **2003**, *23*, 7198–7209. [CrossRef]
47. Urra, H.; Dufey, E.; Lisbona, F.; Rojas-Rivera, D.; Hetz, C. When ER stress reaches a dead end. *Biochim. Biophys. Acta* **2013**, *1833*, 3507–3517. [CrossRef]
48. Marciniak, S.J.; Yun, C.Y.; Oyadomari, S.; Novoa, I.; Zhang, Y.; Jungreis, R.; Nagata, K.; Harding, H.P.; Ron, D. CHOP induces death by promoting protein synthesis and oxidation in the stressed endoplasmic reticulum. *Genes Dev.* **2004**, *18*, 3066–3077. [CrossRef]
49. McCullough, K.D.; Martindale, J.L.; Klotz, L.O.; Aw, T.Y.; Holbrook, N.J. Gadd153 sensitizes cells to endoplasmic reticulum stress by down-regulating Bcl2 and perturbing the cellular redox state. *Mol. Cell Biol.* **2001**, *21*, 1249–1259. [CrossRef]
50. Novoa, I.; Zeng, H.; Harding, H.P.; Ron, D. Feedback inhibition of the unfolded protein response by GADD34-mediated dephosphorylation of eIF2alpha. *J. Cell Biol.* **2001**, *153*, 1011–1022. [CrossRef]
51. Galehdar, Z.; Swan, P.; Fuerth, B.; Callaghan, S.M.; Park, D.S.; Cregan, S.P. Neuronal apoptosis induced by endoplasmic reticulum stress is regulated by ATF4-CHOP-mediated induction of the Bcl-2 homology 3-only member PUMA. *J. Neurosci.* **2010**, *30*, 16938–16948. [CrossRef]
52. Daste, F.; Galli, T.; Tareste, D. Structure and function of longin SNAREs. *J. Cell Sci.* **2015**, *128*, 4263–4272. [CrossRef] [PubMed]
53. Haze, K.; Yoshida, H.; Yanagi, H.; Yura, T.; Mori, K. Mammalian transcription factor ATF6 is synthesized as a transmembrane protein and activated by proteolysis in response to endoplasmic reticulum stress. *Mol. Biol. Cell* **1999**, *10*, 3787–3799. [CrossRef]
54. Yoshida, H. ER stress and diseases. *FEBS J.* **2007**, *274*, 630–658. [CrossRef] [PubMed]
55. Thuerauf, D.J.; Morrison, L.; Glembotski, C.C. Opposing roles for ATF6alpha and ATF6beta in endoplasmic reticulum stress response gene induction. *J. Biol. Chem.* **2004**, *279*, 21078–21084. [CrossRef] [PubMed]
56. Vallejo, M.; Ron, D.; Miller, C.P.; Habener, J.F. C/ATF, a member of the activating transcription factor family of DNA-binding proteins, dimerizes with CAAT/enhancer-binding proteins and directs their binding to cAMP response elements. *Proc. Natl. Acad. Sci. USA* **1993**, *90*, 4679–4683. [CrossRef] [PubMed]
57. Fawcett, T.W.; Martindale, J.L.; Guyton, K.Z.; Hai, T.; Holbrook, N.J. Complexes containing activating transcription factor (ATF)/cAMP-responsive-element-binding protein (CREB) interact with the CCAAT/enhancer-binding protein (C/EBP)-ATF composite site to regulate Gadd153 expression during the stress response. *Biochem. J.* **1999**, *339 Pt 1*, 135–141.

58. Tirasophon, W.; Welihinda, A.A.; Kaufman, R.J. A stress response pathway from the endoplasmic reticulum to the nucleus requires a novel bifunctional protein kinase/endoribonuclease (Ire1p) in mammalian cells. *Genes Dev.* **1998**, *12*, 1812–1824. [CrossRef]
59. Wang, X.Z.; Harding, H.P.; Zhang, Y.; Jolicoeur, E.M.; Kuroda, M.; Ron, D. Cloning of mammalian Ire1 reveals diversity in the ER stress responses. *EMBO J.* **1998**, *17*, 5708–5717. [CrossRef]
60. Iwawaki, T.; Hosoda, A.; Okuda, T.; Kamigori, Y.; Nomura-Furuwatari, C.; Kimata, Y.; Tsuru, A.; Kohno, K. Translational control by the ER transmembrane kinase/ribonuclease IRE1 under ER stress. *Nat. Cell Biol.* **2001**, *3*, 158–164. [CrossRef]
61. Bertolotti, A.; Wang, X.; Novoa, I.; Jungreis, R.; Schlessinger, K.; Cho, J.H.; West, A.B.; Ron, D. Increased sensitivity to dextran sodium sulfate colitis in IRE1beta-deficient mice. *J. Clin. Investig.* **2001**, *107*, 585–593. [CrossRef]
62. Martino, M.B.; Jones, L.; Brighton, B.; Ehre, C.; Abdulah, L.; Davis, C.W.; Ron, D.; O'Neal, W.K.; Ribeiro, C.M. The ER stress transducer IRE1beta is required for airway epithelial mucin production. *Mucosal Immunol.* **2013**, *6*, 639–654. [CrossRef] [PubMed]
63. Prischi, F.; Nowak, P.R.; Carrara, M.; Ali, M.M. Phosphoregulation of Ire1 RNase splicing activity. *Nat. Commun.* **2014**, *5*, 3554. [CrossRef] [PubMed]
64. Yoshida, H.; Matsui, T.; Yamamoto, A.; Okada, T.; Mori, K. XBP1 mRNA is induced by ATF6 and spliced by IRE1 in response to ER stress to produce a highly active transcription factor. *Cell* **2001**, *107*, 881–891. [CrossRef]
65. Wang, Y.; Xing, P.; Cui, W.; Wang, W.; Cui, Y.; Ying, G.; Wang, X.; Li, B. Acute Endoplasmic Reticulum Stress-Independent Unconventional Splicing of XBP1 mRNA in the Nucleus of Mammalian Cells. *Int. J. Mol. Sci.* **2015**, *16*, 13302–13321. [CrossRef] [PubMed]
66. Aragon, T.; van Anken, E.; Pincus, D.; Serafimova, I.M.; Korennykh, A.V.; Rubio, C.A.; Walter, P. Messenger RNA targeting to endoplasmic reticulum stress signalling sites. *Nature* **2009**, *457*, 736–740. [CrossRef] [PubMed]
67. Uemura, A.; Oku, M.; Mori, K.; Yoshida, H. Unconventional splicing of XBP1 mRNA occurs in the cytoplasm during the mammalian unfolded protein response. *J. Cell Sci.* **2009**, *122*, 2877–2886. [CrossRef] [PubMed]
68. Hetz, C. The unfolded protein response: Controlling cell fate decisions under ER stress and beyond. *Nat. Rev. Mol. Cell Biol.* **2012**, *13*, 89–102. [CrossRef]
69. Acosta-Alvear, D.; Zhou, Y.; Blais, A.; Tsikitis, M.; Lents, N.H.; Arias, C.; Lennon, C.J.; Kluger, Y.; Dynlacht, B.D. XBP1 controls diverse cell type- and condition-specific transcriptional regulatory networks. *Mol. Cell* **2007**, *27*, 53–66. [CrossRef]
70. Lee, A.H.; Iwakoshi, N.N.; Glimcher, L.H. XBP-1 regulates a subset of endoplasmic reticulum resident chaperone genes in the unfolded protein response. *Mol. Cell Biol.* **2003**, *23*, 7448–7459. [CrossRef]
71. Hetz, C.; Papa, F.R. The Unfolded Protein Response and Cell Fate Control. *Mol. Cell* **2018**, *69*, 169–181. [CrossRef]
72. Yoshida, H. Cytoplasmic mRNA splicing. *Tanpakushitsu Kakusan Koso* **2006**, *51*, 863–870. [PubMed]
73. Kanda, S.; Yanagitani, K.; Yokota, Y.; Esaki, Y.; Kohno, K. Autonomous translational pausing is required for XBP1u mRNA recruitment to the ER via the SRP pathway. *Proc. Natl. Acad. Sci. USA* **2016**, *113*, E5886–E5895. [CrossRef] [PubMed]
74. Yanagitani, K.; Kimata, Y.; Kadokura, H.; Kohno, K. Translational pausing ensures membrane targeting and cytoplasmic splicing of XBP1u mRNA. *Science* **2011**, *331*, 586–589. [CrossRef] [PubMed]
75. Yanagitani, K.; Imagawa, Y.; Iwawaki, T.; Hosoda, A.; Saito, M.; Kimata, Y.; Kohno, K. Cotranslational targeting of XBP1 protein to the membrane promotes cytoplasmic splicing of its own mRNA. *Mol. Cell* **2009**, *34*, 191–200. [CrossRef] [PubMed]
76. Plumb, R.; Zhang, Z.R.; Appathurai, S.; Mariappan, M. A functional link between the co-translational protein translocation pathway and the UPR. *eLife* **2015**, 4. [CrossRef]
77. Ogata, M.; Hino, S.; Saito, A.; Morikawa, K.; Kondo, S.; Kanemoto, S.; Murakami, T.; Taniguchi, M.; Tanii, I.; Yoshinaga, K.; et al. Autophagy is activated for cell survival after endoplasmic reticulum stress. *Mol. Cell Biol.* **2006**, *26*, 9220–9231. [CrossRef] [PubMed]
78. Yu, L. Reporting the incidence of school violence across grade levels in the U.S. using the Third International Mathematics and Science Study (TIMSS). *J. Appl. Meas.* **2004**, *5*, 287–300. [PubMed]
79. Kim, I.; Shu, C.W.; Xu, W.; Shiau, C.W.; Grant, D.; Vasile, S.; Cosford, N.D.; Reed, J.C. Chemical biology investigation of cell death pathways activated by endoplasmic reticulum stress reveals cytoprotective modulators of ASK1. *J. Biol. Chem.* **2009**, *284*, 1593–1603. [CrossRef]

80. Deng, X.; Xiao, L.; Lang, W.; Gao, F.; Ruvolo, P.; May, W.S., Jr. Novel role for JNK as a stress-activated Bcl2 kinase. *J. Biol. Chem.* **2001**, *276*, 23681–23688. [CrossRef]
81. Puthalakath, H.; O'Reilly, L.A.; Gunn, P.; Lee, L.; Kelly, P.N.; Huntington, N.D.; Hughes, P.D.; Michalak, E.M.; McKimm-Breschkin, J.; Motoyama, N.; et al. ER stress triggers apoptosis by activating BH3-only protein Bim. *Cell* **2007**, *129*, 1337–1349. [CrossRef]
82. Yamaguchi, H.; Wang, H.G. CHOP is involved in endoplasmic reticulum stress-induced apoptosis by enhancing DR5 expression in human carcinoma cells. *J. Biol. Chem.* **2004**, *279*, 45495–45502. [CrossRef] [PubMed]
83. Luo, D.; He, Y.; Zhang, H.; Yu, L.; Chen, H.; Xu, Z.; Tang, S.; Urano, F.; Min, W. AIP1 is critical in transducing IRE1-mediated endoplasmic reticulum stress response. *J. Biol. Chem.* **2008**, *283*, 11905–11912. [CrossRef] [PubMed]
84. Hetz, C.; Bernasconi, P.; Fisher, J.; Lee, A.H.; Bassik, M.C.; Antonsson, B.; Brandt, G.S.; Iwakoshi, N.N.; Schinzel, A.; Glimcher, L.H.; et al. Proapoptotic BAX and BAK modulate the unfolded protein response by a direct interaction with IRE1alpha. *Science* **2006**, *312*, 572–576. [CrossRef]
85. Gu, F.; Nguyen, D.T.; Stuible, M.; Dube, N.; Tremblay, M.L.; Chevet, E. Protein-tyrosine phosphatase 1B potentiates IRE1 signaling during endoplasmic reticulum stress. *J. Biol. Chem.* **2004**, *279*, 49689–49693. [CrossRef]
86. Lisbona, F.; Rojas-Rivera, D.; Thielen, P.; Zamorano, S.; Todd, D.; Martinon, F.; Glavic, A.; Kress, C.; Lin, J.H.; Walter, P.; et al. BAX inhibitor-1 is a negative regulator of the ER stress sensor IRE1alpha. *Mol. Cell* **2009**, *33*, 679–691. [CrossRef] [PubMed]
87. Woehlbier, U.; Hetz, C. Modulating stress responses by the UPRosome: A matter of life and death. *Trends Biochem. Sci.* **2011**, *36*, 329–337. [CrossRef] [PubMed]
88. Hoozemans, J.J.M.; van Haastert, E.S.; Eikelenboom, P.; de Vos, R.A.I.; Rozemuller, J.M.; Scheper, W. Activation of the unfolded protein response in Parkinson's disease. *Biochem. Biophys. Res. Commun.* **2007**, *354*, 707–711. [CrossRef] [PubMed]
89. Baek, J.H.; Whitfield, D.; Howlett, D.; Francis, P.; Bereczki, E.; Ballard, C.; Hortobagyi, T.; Attems, J.; Aarsland, D. Unfolded protein response is activated in Lewy body dementias. *Neuropathol. Appl. Neurobiol.* **2016**, *42*, 352–365. [CrossRef]
90. Baek, J.H.; Mamula, D.; Tingstam, B.; Pereira, M.; He, Y.C.; Svenningsson, P. GRP78 Level Is Altered in the Brain, but Not in Plasma or Cerebrospinal Fluid in Parkinson's Disease Patients. *Front. Neurosci.* **2019**, 13. [CrossRef]
91. Esteves, A.R.; Cardoso, S.M. Differential protein expression in diverse brain areas of Parkinson's and Alzheimer's disease patients. *Sci. Rep.* **2020**, 10. [CrossRef]
92. Selvaraj, S.; Sun, Y.Y.; Watt, J.A.; Wang, S.P.; Lei, S.B.; Birnbaumer, L.; Singh, B.B. Neurotoxin-Induced ER stress in mouse dopaminergic neurons involves downregulation of TRPC1 and inhibition of AKT/mTOR signaling. *J. Clin. Investig.* **2012**, *122*, 1354–1367. [CrossRef] [PubMed]
93. Conn, K.J.; Gao, W.; McKee, A.; Lan, M.S.; Ullman, M.D.; Eisenhauer, P.B.; Fine, R.E.; Wells, J.M. Identification of the protein disulfide isomerase family member PDIp in experimental Parkinson's disease and Lewy body pathology. *Brain Res.* **2004**, *1022*, 164–172. [CrossRef] [PubMed]
94. Freedman, R.B.; Hirst, T.R.; Tuite, M.F. Protein disulphide isomerase: Building bridges in protein folding. *Trends Biochem. Sci.* **1994**, *19*, 331–336. [CrossRef]
95. Turano, C.; Coppari, S.; Altieri, F.; Ferraro, A. Proteins of the PDI family: Unpredicted non-ER locations and functions. *J. Cell. Physiol.* **2002**, *193*, 154–163. [CrossRef] [PubMed]
96. Kaufman, R.J. Stress signaling from the lumen of the endoplasmic reticulum: Coordination of gene transcriptional and translational controls. *Genes Dev.* **1999**, *13*, 1211–1233. [CrossRef] [PubMed]
97. Uehara, T.; Nakamura, T.; Yao, D.D.; Shi, Z.Q.; Gu, Z.Z.; Ma, Y.L.; Masliah, E.; Nomura, Y.; Lipton, S.A. S-Nitrosylated protein-disulphide isomerase links protein misfolding to neurodegeneration. *Nature* **2006**, *441*, 513–517. [CrossRef] [PubMed]
98. Heman-Ackah, S.M.; Manzano, R.; Hoozemans, J.J.M.; Scheper, W.; Flynn, R.; Haerty, W.; Cowley, S.A.; Bassett, A.R.; Wood, M.J.A. Alpha-synuclein induces the unfolded protein response in Parkinson's disease SNCA triplication iPSC-derived neurons. *Hum. Mol. Genet.* **2017**, *26*, 4441–4450. [CrossRef]
99. Slodzinski, H.; Moran, L.B.; Michael, G.J.; Wang, B.; Novoselov, S.; Cheetham, M.E.; Pearce, R.K.B.; Graeber, M.B. Homocysteine-induced endoplasmic reticulum protein (Herp) is up-regulated in parkinsonian substantia nigra and present in the core of Lewy bodies. *Clin. Neuropathol.* **2009**, *28*, 333–343. [CrossRef]
100. Ryu, E.J.; Harding, H.P.; Angelastro, J.M.; Vitolo, O.V.; Ron, D.; Greene, L.A. Endoplasmic reticulum stress and the unfolded protein response in cellular models of Parkinson's disease. *J. Neurosci.* **2002**, *22*, 10690–10698. [CrossRef]

101. Holtz, W.A.; O'Malley, K.L. Parkinsonian mimetics induce aspects of unfolded protein response in death of dopaminergic neurons. *J. Biol. Chem.* **2003**, *278*, 19367–19377. [CrossRef]
102. Hu, L.W.; Yen, J.H.; Shen, Y.T.; Wu, K.Y.; Wu, M.J. Luteolin Modulates 6-Hydroxydopamine-Induced Transcriptional Changes of Stress Response Pathways in PC12 Cells. *PLoS ONE* **2014**, *9*, e97880. [CrossRef] [PubMed]
103. Enogieru, A.B.; Haylett, W.L.; Miller, H.C.; van der Westhuizen, F.H.; Hiss, D.C.; Ekpo, O.E. Attenuation of Endoplasmic Reticulum Stress, Impaired Calcium Homeostasis, and Altered Bioenergetic Functions in MPP+-Exposed SH-SY5Y Cells Pretreated with Rutin. *Neurotox. Res.* **2019**, *36*, 764–776. [CrossRef] [PubMed]
104. Jiao, F.-J.; Wang, Q.-Z.; Zhang, P.; Yan, J.-G.; Zhang, Z.; He, F.; Zhang, Q.; Lv, Z.-X.; Peng, X.; Cai, H.-W.; et al. CDK5-mediated phosphorylation of XBP1s contributes to its nuclear translocation and activation in MPP+-induced Parkinson's disease model. *Sci. Rep.* **2017**, 7. [CrossRef] [PubMed]
105. Xie, L.; Tiong, C.X.; Bian, J.S. Hydrogen sulfide protects SH-SY5Y cells against 6-hydroxydopamine-induced endoplasmic reticulum stress. *Am. J. Physiol. Cell Physiol.* **2012**, *303*, C81–C91. [CrossRef]
106. Deng, C.; Tao, R.; Yu, S.Z.; Jin, H. Inhibition of 6-hydroxydopamine-induced endoplasmic reticulum stress by sulforaphane through the activation of Nrf2 nuclear translocation. *Mol. Med. Rep.* **2012**, *6*, 215–219. [CrossRef]
107. Yamamuro, A.; Yoshioka, Y.; Ogita, K.; Maeda, S. Involvement of endoplasmic reticulum stress on the cell death induced by 6-hydroxydopamine in human neuroblastoma SH-SY5Y cells. *Neurochem. Res.* **2006**, *31*, 657–664. [CrossRef]
108. Yang, W.; Tiffany-Castiglioni, E.; Koh, H.C.; Son, I.H. Paraquat activates the IRE1/ASK1/JNK cascade associated with apoptosis in human neuroblastoma SH-SY5Y cells. *Toxicol. Lett.* **2009**, *191*, 203–210. [CrossRef]
109. Chinta, S.J.; Rane, A.; Poksay, K.S.; Bredesen, D.E.; Andersen, J.K.; Rao, R.V. Coupling Endoplasmic Reticulum Stress to the Cell Death Program in Dopaminergic Cells: Effect of Paraquat. *Neuromol. Med.* **2008**, *10*, 333–342. [CrossRef]
110. Tong, Q.; Wu, L.; Gao, Q.; Ou, Z.; Zhu, D.Y.; Zhang, Y.D. PPAR beta/delta Agonist Provides Neuroprotection by Suppression of IRE1 alpha-Caspase-12-Mediated Endoplasmic Reticulum Stress Pathway in the Rotenone Rat Model of Parkinson's Disease. *Mol. Neurobiol.* **2016**, *53*, 3822–3831. [CrossRef]
111. Tong, Q.; Wu, L.; Jiang, T.; Ou, Z.; Zhang, Y.D.; Zhu, D.Y. Inhibition of endoplasmic reticulum stress-activated IRE1 alpha-TRAF2-caspase-12 apoptotic pathway is involved in the neuroprotective effects of telmisartan in the rotenone rat model of Parkinson's disease. *Eur. J. Pharmacol.* **2016**, *776*, 106–115. [CrossRef]
112. Ghribi, O.; Herman, M.M.; Pramoonjago, P.; Savory, J. MPP+ induces the endoplasmic reticulum stress response in rabbit brain involving activation of the ATF-6 and NF-kappaB signaling pathways. *J. Neuropathol. Exp. Neurol.* **2003**, *62*, 1144–1153. [CrossRef] [PubMed]
113. Cai, P.; Ye, J.; Zhu, J.; Liu, D.; Chen, D.; Wei, X.; Johnson, N.R.; Wang, Z.; Zhang, H.; Cao, G.; et al. Inhibition of Endoplasmic Reticulum Stress is Involved in the Neuroprotective Effect of bFGF in the 6-OHDA-Induced Parkinson's Disease Model. *Aging Dis.* **2016**, *7*, 336–449. [CrossRef] [PubMed]
114. Coppola-Segovia, V.; Cavarsan, C.; Maia, F.G.; Ferraz, A.C.; Nakao, L.S.; Lima, M.M.S.; Zanata, S.M. ER Stress Induced by Tunicamycin Triggers alpha-Synuclein Oligomerization, Dopaminergic Neurons Death and Locomotor Impairment: A New Model of Parkinson's Disease. *Mol. Neurobiol.* **2017**, *54*, 5798–5806. [CrossRef] [PubMed]
115. Alves da Costa, C. Recent advances on alpha-synuclein cell biology: Functions and dysfunctions. *Curr. Mol. Med.* **2003**, *3*, 17–24. [CrossRef]
116. Surguchov, A. Molecular and cellular biology of synucleins. *Int. Rev. Cell Mol. Biol.* **2008**, *270*, 225–317. [CrossRef]
117. Roberts, H.L.; Brown, D.R. Seeking a mechanism for the toxicity of oligomeric alpha-synuclein. *Biomolecules* **2015**, *5*, 282–305. [CrossRef]
118. Alam, P.; Bousset, L.; Melki, R.; Otzen, D.E. Alpha-synuclein oligomers and fibrils: A spectrum of species, a spectrum of toxicities. *J. Neurochem.* **2019**, *150*, 522–534. [CrossRef]
119. Cooper, A.A.; Gitler, A.D.; Cashikar, A.; Haynes, C.M.; Hill, K.J.; Bhullar, B.; Liu, K.; Xu, K.; Strathearn, K.E.; Liu, F.; et al. Alpha-synuclein blocks ER-Golgi traffic and Rab1 rescues neuron loss in Parkinson's models. *Science* **2006**, *313*, 324–328. [CrossRef]
120. Sugeno, N.; Takeda, A.; Hasegawa, T.; Kobayashi, M.; Kikuchi, A.; Mori, F.; Wakabayashi, K.; Itoyama, Y. Serine 129 phosphorylation of alpha-synuclein induces unfolded protein response-mediated cell death. *J. Biol. Chem.* **2008**, *283*, 23179–23188. [CrossRef]

121. Shin, J.K.; Ki, S.S.; Yunhee, K.K. α-Synuclein Induces Unfolded Protein Response Via Distinct Signaling Pathway Independent of ER-membrane Kinases. *Integr. Biosci.* **2006**, *10*, 115–120.
122. Castillo-Carranza, D.L.; Zhang, Y.; Guerrero-Munoz, M.J.; Kayed, R.; Rincon-Limas, D.E.; Fernandez-Funez, P. Differential Activation of the ER Stress Factor XBP1 by Oligomeric Assemblies. *Neurochem. Res.* **2012**, *37*, 1707–1717. [CrossRef] [PubMed]
123. Jiang, P.; Gan, M.; Ebrahim, A.S.; Lin, W.-L.; Melrose, H.L.; Yen, S.-H.C. ER stress response plays an important role in aggregation of alpha-synuclein. *Mol. Neurodegener.* **2010**, 5. [CrossRef]
124. Bellucci, A.; Navarria, L.; Zaltieri, M.; Falarti, E.; Bodei, S.; Sigala, S.; Battistin, L.; Spillantini, M.; Missale, C.; Spano, P. Induction of the unfolded protein response by alpha-synuclein in experimental models of Parkinson's disease. *J. Neurochem.* **2011**, *116*, 588–605. [CrossRef] [PubMed]
125. Credle, J.J.; Forcelli, P.A.; Delannoy, M.; Oaks, A.W.; Permaul, E.; Berry, D.L.; Duka, V.; Wills, J.; Sidhu, A. Alpha-Synuclein-mediated inhibition of ATF6 processing into COPII vesicles disrupts UPR signaling in Parkinson's disease. *Neurobiol. Dis.* **2015**, *76*, 112–125. [CrossRef]
126. Paiva, I.; Jain, G.; Lazaro, D.F.; Jercic, K.G.; Hentrich, T.; Kerimoglu, C.; Pinho, R.; Szego, E.M.; Burkhardt, S.; Capece, V.; et al. Alpha-synuclein deregulates the expression of COL4A2 and impairs ER-Golgi function. *Neurobiol. Dis.* **2018**, *119*, 121–135. [CrossRef]
127. Smith, W.W.; Jiang, H.; Pei, Z.; Tanaka, Y.; Morita, H.; Sawa, A.; Dawson, V.L.; Dawson, T.M.; Ross, C.A. Endoplasmic reticulum stress and mitochondrial cell death pathways mediate A53T mutant alpha-synuclein-induced toxicity. *Hum. Mol. Genet.* **2005**, *14*, 3801–3811. [CrossRef]
128. Liu, M.C.; Yu, S.T.; Wang, J.W.; Qiao, J.H.; Liu, Y.; Wang, S.M.; Zhao, Y. Ginseng protein protects against mitochondrial dysfunction and neurodegeneration by inducing mitochondrial unfolded protein response in Drosophila melanogaster PINK1 model of Parkinson's disease. *J. Ethnopharmacol.* **2020**, 247. [CrossRef]
129. Colla, E.; Coune, P.; Liu, Y.; Pletnikova, O.; Troncoso, J.C.; Iwatsubo, T.; Schneider, B.L.; Lee, M.K. Endoplasmic Reticulum Stress Is Important for the Manifestations of alpha-Synucleinopathy In Vivo. *J. Neurosci.* **2012**, *32*, 3306–3320. [CrossRef]
130. Colla, E.; Schneider, B.; Coune, P.; Jensen, P.H.; Troncoso, J.C.; Lee, M.K. Toxic alpha-synuclein oligomer accumulation and endoplasmic reticulum stress is mechanistically linked to alpha-synucleinopathy in vivo. *Mol. Biol. Cell* **2012**, *23*, 8120.
131. Miraglia, F.; Valvano, V.; Rota, L.; Di Primio, C.; Quercioli, V.; Betti, L.; Giannaccini, G.; Cattaneo, A.; Colla, E. Alpha-Synuclein FRET Biosensors Reveal Early Alpha-Synuclein Aggregation in the Endoplasmic Reticulum. *Life* **2020**, *10*, 147. [CrossRef]
132. Bosgraaf, L.; Van Haastert, P.J. Roc, a Ras/GTPase domain in complex proteins. *Biochim. Biophys. Acta* **2003**, *1643*, 5–10. [CrossRef] [PubMed]
133. Paisan-Ruiz, C.; Jain, S.; Evans, E.W.; Gilks, W.P.; Simon, J.; van der Brug, M.; Lopez de Munain, A.; Aparicio, S.; Gil, A.M.; Khan, N.; et al. Cloning of the gene containing mutations that cause PARK8-linked Parkinson's disease. *Neuron* **2004**, *44*, 595–600. [CrossRef] [PubMed]
134. Zimprich, A.; Biskup, S.; Leitner, P.; Lichtner, P.; Farrer, M.; Lincoln, S.; Kachergus, J.; Hulihan, M.; Uitti, R.J.; Calne, D.B.; et al. Mutations in LRRK2 cause autosomal-dominant parkinsonism with pleomorphic pathology. *Neuron* **2004**, *44*, 601–607. [CrossRef] [PubMed]
135. Paisan-Ruiz, C.; Nath, P.; Washecka, N.; Gibbs, J.R.; Singleton, A.B. Comprehensive analysis of LRRK2 in publicly available Parkinson's disease cases and neurologically normal controls. *Hum. Mutat.* **2008**, *29*, 485–490. [CrossRef]
136. Vitte, J.; Traver, S.; De Paula, A.M.; Lesage, S.; Rovelli, G.; Corti, O.; Duyckaerts, C.; Brice, A. Leucine-Rich Repeat Kinase 2 Is Associated With the Endoplasmic Reticulum in Dopaminergic Neurons and Accumulates in the Core of Lewy Bodies in Parkinson Disease. *J. Neuropathol. Exp. Neurol.* **2010**, *69*, 959–972. [CrossRef]
137. Yuan, Y.Y.; Cao, P.X.; Smith, M.A.; Kramp, K.; Huang, Y.; Hisamoto, N.; Matsumoto, K.; Hatzoglou, M.; Jin, H.; Feng, Z.Y. Dysregulated LRRK2 Signaling in Response to Endoplasmic Reticulum Stress Leads to Dopaminergic Neuron Degeneration in C. elegans. *PLoS ONE* **2011**, 6. [CrossRef]
138. Toyofuku, T.; Okamoto, Y.; Ishikawa, T.; Sasawatari, S.; Kumanogoh, A. LRRK2 regulates endoplasmic reticulum-mitochondrial tethering through the PERK-mediated ubiquitination pathway. *EMBO J.* **2019**. [CrossRef]
139. Lee, J.H.; Han, J.-H.; Kim, H.; Park, S.M.; Joe, E.-H.; Jou, I. Parkinson's disease-associated LRRK2-G2019S mutant acts through regulation of SERCA activity to control ER stress in astrocytes. *Acta Neuropathol. Commun.* **2019**, 7. [CrossRef]

140. Shimura, H.; Hattori, N.; Kubo, S.; Mizuno, Y.; Asakawa, S.; Minoshima, S.; Shimizu, N.; Iwai, K.; Chiba, T.; Tanaka, K.; et al. Familial Parkinson disease gene product, parkin, is a ubiquitin-protein ligase. *Nat. Genet.* **2000**, *25*, 302–305. [CrossRef]
141. Imai, Y.; Soda, M.; Takahashi, R. Parkin suppresses unfolded protein stress-induced cell death through its E3 ubiquitin-protein ligase activity. *J. Biol. Chem.* **2000**, *275*, 35661–35664. [CrossRef]
142. Da Costa, C.; Sunyach, C.; Giaime, E.; West, A.; Corti, O.; Brice, A.; Safe, S.; Abou-Sleiman, P.; Wood, N.; Takahashi, H.; et al. Transcriptional repression of p53 by parkin and impairment by mutations associated with autosomal recessive juvenile Parkinson's disease. *Nat. Cell Biol.* **2009**, *11*, 1370–1375. [CrossRef] [PubMed]
143. Imai, Y.; Soda, M.; Inoue, H.; Hattori, N.; Mizuno, Y.; Takahashi, R. An unfolded putative transmembrane polypeptide, which can lead to endoplasmic reticulum stress, is a substrate of parkin. *Cell* **2001**, *105*, 891–902. [CrossRef]
144. Bouman, L.; Schlierf, A.; Lutz, A.K.; Shan, J.; Deinlein, A.; Kast, J.; Galehdar, Z.; Palmisano, V.; Patenge, N.; Berg, D.; et al. Parkin is transcriptionally regulated by ATF4: Evidence for an interconnection between mitochondrial stress and ER stress. *Cell Death Differ.* **2011**, *18*, 769–782. [CrossRef] [PubMed]
145. Wu, L.; Luo, N.; Zhao, H.R.; Gao, Q.; Lu, J.; Pan, Y.; Shi, J.P.; Tian, Y.Y.; Zhang, Y.D. Salubrinal protects against rotenone-induced SH-SY5Y cell death via ATF4-parkin pathway. *Brain Res.* **2014**, *1549*, 52–62. [CrossRef] [PubMed]
146. Cali, T.; Ottolini, D.; Brini, M. Mitochondria, calcium, and endoplasmic reticulum stress in Parkinson's disease. *Biofactors* **2011**, *37*, 228–240. [CrossRef] [PubMed]
147. Duplan, E.; Giaime, E.; Viotti, J.; Sévalle, J.; Corti, O.; Brice, A.; Ariga, H.; Qi, L.; Checler, F.; Alves da Costa, C. ER-stress-associated functional link between Parkin and DJ-1 via a transcriptional cascade involving the tumor suppressor p53 and the spliced X-box binding protein XBP-1. *J. Cell Sci.* **2013**, *126*, 2124–2133. [CrossRef] [PubMed]
148. Singh, K.; Han, K.; Tilve, S.; Wu, K.; Geller, H.M.; Sack, M.N. Parkin targets NOD2 to regulate astrocyte endoplasmic reticulum stress and inflammation. *Glia* **2018**, *66*, 2427–2437. [CrossRef] [PubMed]
149. Ledesma, M.D.; Galvan, C.; Hellias, B.; Dotti, C.; Jensen, P.H. Astrocytic but not neuronal increased expression and redistribution of parkin during unfolded protein stress. *J. Neurochem.* **2002**, *83*, 1431–1440. [CrossRef]
150. Wang, H.Q.; Imai, Y.; Kataoka, A.; Takahashi, R. Forum original research communication—Cell type-specific upregulation of parkin in response to ER stress. *Antioxid. Redox Signal.* **2007**, *9*, 533–541. [CrossRef]
151. Celardo, I.; Costa, A.C.; Lehmann, S.; Jones, C.; Wood, N.; Mencacci, N.E.; Mallucci, G.R.; Loh, S.H.Y.; Martins, L.M. Mitofusin-mediated ER stress triggers neurodegeneration in pink1/parkin models of Parkinson's disease. *Cell Death Dis.* **2016**, 7. [CrossRef]
152. Rana, A.; Rera, M.; Walker, D.W. Parkin overexpression during aging reduces proteotoxicity, alters mitochondrial dynamics, and extends lifespan. *Proc. Natl. Acad. Sci. USA* **2013**, *110*, 8638–8643. [CrossRef] [PubMed]
153. Gao, F.; Yang, J.; Wang, D.; Li, C.; Fu, Y.; Wang, H.; He, W.; Zhang, J. Mitophagy in Parkinson's Disease: Pathogenic and Therapeutic Implications. *Front. Neurol.* **2017**, 8. [CrossRef] [PubMed]
154. Youle, R.J.; Narendra, D.P. Mechanisms of mitophagy. *Nat. Rev. Mol. Cell Biol.* **2011**, *12*, 9–14. [CrossRef] [PubMed]
155. Valente, E.M.; Abou-Sleiman, P.M.; Caputo, V.; Muqit, M.M.; Harvey, K.; Gispert, S.; Ali, Z.; Del Turco, D.; Bentivoglio, A.R.; Healy, D.G.; et al. Hereditary early-onset Parkinson's disease caused by mutations in PINK1. *Science* **2004**, *304*, 1158–1160. [CrossRef] [PubMed]
156. Valente, E.M.; Salvi, S.; Ialongo, T.; Marongiu, R.; Elia, A.E.; Caputo, V.; Romito, L.; Albanese, A.; Dallapiccola, B.; Bentivoglio, A.R. PINK1 mutations are associated with sporadic early-onset parkinsonism. *Ann. Neurol.* **2004**, *56*, 336–341. [CrossRef]
157. Zhao, Q.; Wang, J.; Levichkin, I.V.; Stasinopoulos, S.; Ryan, M.T.; Hoogenraad, N.J. A mitochondrial specific stress response in mammalian cells. *EMBO J.* **2002**, *21*, 4411–4419. [CrossRef]
158. Jin, S.M.; Youle, R.J. The accumulation of misfolded proteins in the mitochondrial matrix is sensed by PINK1 to induce PARK2/Parkin-mediated mitophagy of polarized mitochondria. *Autophagy* **2013**, *9*, 1750–1757. [CrossRef]
159. De Castro, P.; Costa, A.C.; Lam, D.; Tufi, R.; Fedele, V.; Moisoi, N.; Dinsdale, D.; Deas, E.; Loh, S.H.Y.; Martins, L.M. Genetic analysis of mitochondrial protein misfolding in Drosophila melanogaster. *Cell Death Differ.* **2012**, *19*, 1308–1316. [CrossRef]
160. Li, L.; Hu, G.K. Pink1 protects cortical neurons from thapsigargin-induced oxidative stress and neuronal apoptosis. *Biosci. Rep.* **2015**, 35. [CrossRef]

161. Torres-Odio, S.; Key, J.; Hoepken, H.H.; Canet-Pons, J.; Valek, L.; Roller, B.; Walter, M.; Morales-Gordo, B.; Meierhofer, D.; Harter, P.N.; et al. Progression of pathology in PINK1-deficient mouse brain from splicing via ubiquitination, ER stress, and mitophagy changes to neuroinflammation. *J. Neuroinflamm.* **2017**, *14*, 154. [CrossRef]
162. Da Costa, C.A. DJ-1: A newcomer in Parkinson's disease pathology. *Curr. Mol. Med.* **2007**, *7*, 650–657. [CrossRef] [PubMed]
163. Lee, S.J.; Kim, S.J.; Kim, I.K.; Ko, J.; Jeong, C.S.; Kim, G.H.; Park, C.; Kang, S.O.; Suh, P.G.; Lee, H.S.; et al. Crystal structures of human DJ-1 and Escherichia coli Hsp31, which share an evolutionarily conserved domain. *J. Biol. Chem.* **2003**, *278*, 44552–44559. [CrossRef] [PubMed]
164. Parrado-Fernandez, C.; Schneider, B.; Ankarcrona, M.; Conti, M.M.; Cookson, M.R.; Kivipelto, M.; Cedazo-Minguez, A.; Sandebring-Matton, A. Reduction of PINK1 or DJ-1 impair mitochondrial motility in neurites and alter ER-mitochondria contacts. *J. Cell Mol. Med.* **2018**, *22*, 5439–5449. [CrossRef] [PubMed]
165. Ottolini, D.; Calì, T.; Negro, A.; Brini, M. The Parkinson disease-related protein DJ-1 counteracts mitochondrial impairment induced by the tumour suppressor protein p53 by enhancing endoplasmic reticulum-mitochondria tethering. *Hum. Mol. Genet.* **2013**. [CrossRef]
166. Yokota, T.; Sugawara, K.; Ito, K.; Takahashi, R.; Ariga, H.; Mizusawa, H. Down regulation of DJ-1 enhances cell death by oxidative stress, ER stress, and proteasome inhibition. *Biochem. Biophys. Res. Commun.* **2003**, *312*, 1342–1348. [CrossRef]
167. Yang, J.; Kim, K.S.; Iyirhiaro, G.O.; Marcogliese, P.C.; Callaghan, S.M.; Qu, D.B.; Kim, W.J.; Slack, R.S.; Park, D.S. DJ-1 modulates the unfolded protein response and cell death via upregulation of ATF4 following ER stress. *Cell Death. Dis.* **2019**, 10. [CrossRef]
168. Lee, D.H.; Kim, D.; Kim, S.T.; Jeong, S.; Kim, J.L.; Shim, S.M.; Heo, A.J.; Song, X.; Guo, Z.S.; Bartlett, D.L.; et al. PARK7 modulates autophagic proteolysis through binding to the N-terminally arginylated form of the molecular chaperone HSPA5. *Autophagy* **2018**, *14*, 1870–1885. [CrossRef]
169. Krebs, C.E.; Karkheiran, S.; Powell, J.C.; Cao, M.; Makarov, V.; Darvish, H.; Di Paolo, G.; Walker, R.H.; Shahidi, G.A.; Buxbaum, J.D.; et al. The Sac1 domain of SYNJ1 identified mutated in a family with early-onset progressive Parkinsonism with generalized seizures. *Hum. Mutat.* **2013**, *34*, 1200–1207. [CrossRef]
170. Quadri, M.; Fang, M.; Picillo, M.; Olgiati, S.; Breedveld, G.J.; Graafland, J.; Wu, B.; Xu, F.; Erro, R.; Amboni, M.; et al. Mutation in the SYNJ1 gene associated with autosomal recessive, early-onset Parkinsonism. *Hum. Mutat.* **2013**, *34*, 1208–1215. [CrossRef]
171. Amodio, G.; Moltedo, O.; Fasano, D.; Zerillo, L.; Oliveti, M.; Di Pietro, P.; Faraonio, R.; Barone, P.; Pellecchia, M.T.; De Rose, A.; et al. PERK-Mediated Unfolded Protein Response Activation and Oxidative Stress in PARK20 Fibroblasts. *Front. Neurosci.* **2019**, 13. [CrossRef]
172. Paisan-Ruiz, C.; Bhatia, K.P.; Li, A.; Hernandez, D.; Davis, M.; Wood, N.W.; Hardy, J.; Houlden, H.; Singleton, A.; Schneider, S.A. Characterization of PLA2G6 as a locus for dystonia-parkinsonism. *Ann. Neurol.* **2009**, *65*, 19–23. [CrossRef]
173. Chiu, C.C.; Lu, C.S.; Weng, Y.H.; Chen, Y.L.; Huang, Y.Z.; Chen, R.S.; Cheng, Y.C.; Huang, Y.C.; Liu, Y.C.; Lai, S.C.; et al. PARK14 (D331Y) PLA2G6 Causes Early-Onset Degeneration of Substantia Nigra Dopaminergic Neurons by Inducing Mitochondrial Dysfunction, ER Stress, Mitophagy Impairment and Transcriptional Dysregulation in a Knockin Mouse Model. *Mol. Neurobiol.* **2019**, *56*, 3835–3853. [CrossRef] [PubMed]
174. Sidransky, E.; Samaddar, T.; Tayebi, N. Mutations in GBA are associated with familial Parkinson disease susceptibility and age at onset. *Neurology* **2009**, *73*, 1424–1425, author reply 1425-1426. [CrossRef] [PubMed]
175. Stojkovska, I.; Krainc, D.; Mazzulli, J.R. Molecular mechanisms of alpha-synuclein and GBA1 in Parkinson's disease. *Cell Tissue Res.* **2018**, *373*, 51–60. [CrossRef]
176. Kurzawa-Akanbi, M.; Hanson, P.S.; Blain, P.G.; Lett, D.J.; McKeith, I.G.; Chinnery, P.F.; Morris, C.M. Glucocerebrosidase Mutations alter the endoplasmic reticulum and lysosomes in Lewy body disease. *J. Neurochem.* **2012**, *123*, 298–309. [CrossRef] [PubMed]
177. Maor, G.; Rencus-Lazar, S.; Filocamo, M.; Steller, H.; Segal, D.; Horowitz, M. Unfolded protein response in Gaucher disease: From human to Drosophila. *Orphanet J. Rare Dis.* **2013**, *8*. [CrossRef] [PubMed]
178. Maor, G.; Cabasso, O.; Krivoruk, O.; Rodriguez, J.; Steller, H.; Segal, D.; Horowitz, M. The contribution of mutant GBA to the development of Parkinson disease in Drosophila. *Hum. Mol. Genet.* **2016**, *25*, 2712–2727. [CrossRef]
179. Strauss, K.M.; Martins, L.M.; Plun-Favreau, H.; Marx, F.P.; Kautzmann, S.; Berg, D.; Gasser, T.; Wszolek, Z.; Muller, T.; Bornemann, A.; et al. Loss of function mutations in the gene encoding Omi/HtrA2 in Parkinson's disease. *Hum. Mol. Genet.* **2005**, *14*, 2099–2111. [CrossRef]

180. Tran, J.; Anastacio, H.; Bardy, C. Genetic predispositions of Parkinson's disease revealed in patient-derived brain cells. *NPJ Parkinsons Dis.* **2020**, *6*, 8. [CrossRef]
181. Luo, F.; Wei, L.; Sun, C.; Chen, X.; Wang, T.; Li, Y.; Liu, Z.; Chen, Z.; Xu, P. HtrA2/Omi is involved in 6-OHDA-induced endoplasmic reticulum stress in SH-SY5Y cells. *J. Mol. Neurosci.* **2012**, *47*, 120–127. [CrossRef]
182. Moisoi, N.; Klupsch, K.; Fedele, V.; East, P.; Sharma, S.; Renton, A.; Plun-Favreau, H.; Edwards, R.E.; Teismann, P.; Esposti, M.D.; et al. Mitochondrial dysfunction triggered by loss of HtrA2 results in the activation of a brain-specific transcriptional stress response. *Cell Death Differ.* **2009**, *16*, 449–464. [CrossRef] [PubMed]
183. Han, C.; Nam, M.K.; Park, H.J.; Seong, Y.M.; Kang, S.; Rhim, H. Tunicamycin-induced ER stress upregulates the expression of mitochondrial HtrA2 and promotes apoptosis through the cytosolic release of HtrA2. *J. Microbiol. Biotechnol.* **2008**, *18*, 1197–1202. [PubMed]
184. Gully, J.C.; Sergeyev, V.G.; Bhootada, Y.; Mendez-Gomez, H.; Meyers, C.A.; Zolotukhin, S.; Gorbatyuk, M.S.; Gorbatyuk, O.S. Up-regulation of activating transcription factor 4 induces severe loss of dopamine nigral neurons in a rat model of Parkinson's disease. *Neurosci. Lett.* **2016**, *627*, 36–41. [CrossRef] [PubMed]
185. Valdes, P.; Mercado, G.; Vidal, R.L.; Molina, C.; Parsons, G.; Court, F.A.; Martinez, A.; Galleguillos, D.; Armentano, D.; Schneider, B.L.; et al. Control of dopaminergic neuron survival by the unfolded protein response transcription factor XBP1. *Proc. Natl. Acad. Sci. USA* **2014**, *111*, 6804–6809. [CrossRef] [PubMed]
186. Sado, M.; Yamasaki, Y.; Iwanaga, T.; Onaka, Y.; Ibuki, T.; Nishihara, S.; Mizuguchi, H.; Momota, H.; Kishibuchi, R.; Hashimoto, T.; et al. Protective effect against Parkinson's disease-related insults through the activation of XBP1. *Brain Res.* **2009**, *1257*, 16–24. [CrossRef] [PubMed]
187. Mattson, M.P. Hormesis defined. *Ageing Res. Rev.* **2008**, *7*, 1–7. [CrossRef] [PubMed]
188. Rzechorzek, N.M.; Connick, P.; Patani, R.; Selvaraj, B.T.; Chandran, S. Hypothermic Preconditioning of Human Cortical Neurons Requires Proteostatic Priming. *EBioMedicine* **2015**, *2*, 528–535. [CrossRef]
189. Mollereau, B.; Manie, S.; Napoletano, F. Getting the better of ER stress. *J. Cell Commun. Signal.* **2014**, *8*, 311–321. [CrossRef]
190. Mollereau, B.; Rzechorzek, N.M.; Roussel, B.D.; Sedru, M.; Van den Brink, D.M.; Bailly-Maitre, B.; Palladino, F.; Medinas, D.B.; Domingos, P.M.; Hunot, S.; et al. Adaptive preconditioning in neurological diseases—Therapeutic insights from proteostatic perturbations. *Brain Res.* **2016**, *1648*, 603–616. [CrossRef]
191. Fouillet, A.; Levet, C.; Virgone, A.; Robin, M.; Dourlen, P.; Rieusset, J.; Belaidi, E.; Ovize, M.; Touret, M.; Nataf, S.; et al. ER stress inhibits neuronal death by promoting autophagy. *Autophagy* **2012**, *8*, 915–926. [CrossRef]
192. Kosmaczewski, S.G.; Edwards, T.J.; Han, S.M.; Eckwahl, M.J.; Meyer, B.I.; Peach, S.; Hesselberth, J.R.; Wolin, S.L.; Hammarlund, M. The RtcB RNA ligase is an essential component of the metazoan unfolded protein response. *EMBO Rep.* **2014**, *15*, 1278–1285. [CrossRef] [PubMed]
193. Ray, A.; Zhang, S.; Rentas, C.; Caldwell, K.A.; Caldwell, G.A. RTCB-1 mediates neuroprotection via XBP-1 mRNA splicing in the unfolded protein response pathway. *J. Neurosci.* **2014**, *34*, 16076–16085. [CrossRef] [PubMed]
194. Si, L.; Xu, T.; Wang, F.; Liu, Q.; Cui, M. X-box-binding protein 1-modified neural stem cells for treatment of Parkinson's disease. *Neural Regen. Res.* **2012**, *7*, 736–740. [CrossRef] [PubMed]
195. Gorbatyuk, M.S.; Shabashvili, A.; Chen, W.; Meyers, C.; Sullivan, L.F.; Salganik, M.; Lin, J.H.; Lewin, A.S.; Muzyczka, N.; Gorbatyuk, O.S. Glucose regulated protein 78 diminishes alpha-synuclein neurotoxicity in a rat model of Parkinson disease. *Mol. Ther.* **2012**, *20*, 1327–1337. [CrossRef]
196. Salganik, M.; Sergeyev, V.G.; Shinde, V.; Meyers, C.A.; Gorbatyuk, M.S.; Lin, J.H.; Zolotukhin, S.; Gorbatyuk, O.S. The loss of glucose-regulated protein 78 (GRP78) during normal aging or from siRNA knockdown augments human alpha-synuclein (alpha-syn) toxicity to rat nigral neurons. *Neurobiol. Aging* **2015**, *36*, 2213–2223. [CrossRef]
197. Omura, T.; Kaneko, M.; Okuma, Y.; Orba, Y.; Nagashima, K.; Takahashi, R.; Fujitani, N.; Matsumura, S.; Hata, A.; Kubota, K.; et al. A ubiquitin ligase HRD1 promotes the degradation of Pael receptor, a substrate of Parkin. *J. Neurochem.* **2006**, *99*, 1456–1469. [CrossRef]
198. Silva, R.M.; Ries, V.; Oo, T.F.; Yarygina, O.; Jackson-Lewis, V.; Ryu, E.J.; Lu, P.D.; Marciniak, S.J.; Ron, D.; Przedborski, S.; et al. CHOP/GADD153 is a mediator of apoptotic death in substantia nigra dopamine neurons in an in vivo neurotoxin model of parkinsonism. *J. Neurochem.* **2005**, *95*, 974–986. [CrossRef]
199. Hashida, K.; Kitao, Y.; Sudo, H.; Awa, Y.; Maeda, S.; Mori, K.; Takahashi, R.; Iinuma, M.; Hori, O. ATF6alpha Promotes Astroglial Activation and Neuronal Survival in a Chronic Mouse Model of Parkinson's Disease. *PLoS ONE* **2012**, *7*, e47950. [CrossRef]

200. Bernales, S.; Morales Soto, M.; McCullagh, E. Unfolded protein stress in the endoplasmic reticulum and mitochondria: A role in neurodegeneration. *Front. Aging Neurosci.* **2012**, 4. [CrossRef]
201. Ji, T.; Zhang, X.; Xin, Z.; Xu, B.; Jin, Z.; Wu, J.; Hu, W.; Yang, Y. Does perturbation in the mitochondrial protein folding pave the way for neurodegeneration diseases? *Ageing Res. Rev.* **2020**, *57*, 100997. [CrossRef]
202. Devi, L.; Raghavendran, V.; Prabhu, B.M.; Avadhani, N.G.; Anandatheerthavarada, H.K. Mitochondrial import and accumulation of alpha-synuclein impair complex I in human dopaminergic neuronal cultures and Parkinson disease brain. *J. Biol. Chem.* **2008**, *283*, 9089–9100. [CrossRef] [PubMed]
203. Shavali, S.; Brown-Borg, H.M.; Ebadi, M.; Porter, J. Mitochondrial localization of alpha-synuclein protein in alpha-synuclein overexpressing cells. *Neurosci. Lett.* **2008**, *439*, 125–128. [CrossRef] [PubMed]
204. Pridgeon, J.W.; Olzmann, J.A.; Chin, L.S.; Li, L. PINK1 Protects against Oxidative Stress by Phosphorylating Mitochondrial Chaperone TRAP1. *PLoS Biol.* **2007**, *5*, e172. [CrossRef] [PubMed]
205. Plun-Favreau, H.; Klupsch, K.; Moisoi, N.; Gandhi, S.; Kjaer, S.; Frith, D.; Harvey, K.; Deas, E.; Harvey, R.J.; McDonald, N.; et al. The mitochondrial protease HtrA2 is regulated by Parkinson's disease-associated kinase PINK1. *Nat. Cell Biol.* **2007**, *9*, 1243–1252. [CrossRef] [PubMed]
206. Rakovic, A.; Grunewald, A.; Voges, L.; Hofmann, S.; Orolicki, S.; Lohmann, K.; Klein, C. PINK1-Interacting Proteins: Proteomic Analysis of Overexpressed PINK1. *Parkinsons Dis.* **2011**, *2011*, 153979. [CrossRef] [PubMed]
207. Martinez, A.; Lopez, N.; Gonzalez, C.; Hetz, C. Targeting of the unfolded protein response (UPR) as therapy for Parkinson's disease. *Biol. Cell* **2019**. [CrossRef]
208. Enogieru, A.B.; Omoruyi, S.I.; Hiss, D.C.; Ekpo, O.E. GRP78/BIP/HSPA5 as a Therapeutic Target in Models of Parkinson's Disease: A Mini Review. *Adv. Pharmacol. Sci.* **2019**. [CrossRef]
209. Mercado, G.; Castillo, V.; Soto, P.; Lopez, N.; Axten, J.M.; Sardi, S.P.; Hoozemans, J.J.M.; Hetz, C. Targeting PERK signaling with the small molecule GSK2606414 prevents neurodegeneration in a model of Parkinson's disease. *Neurobiol. Dis.* **2018**, *112*, 136–148. [CrossRef]

Publisher's Note: MDPI stays neutral with regard to jurisdictional claims in published maps and institutional affiliations.

 *cells*

Review

Alpha-Synuclein: Mechanisms of Release and Pathology Progression in Synucleinopathies

Inês C. Brás [1] and Tiago F. Outeiro [1,2,3,4,*]

1 Center for Biostructural Imaging of Neurodegeneration, Department of Experimental Neurodegeneration, University Medical Center Göttingen, 37075 Göttingen, Germany; ibras@gwdg.de
2 Max Planck Institute for Experimental Medicine, 37075 Göttingen, Germany
3 Faculty of Medical Sciences, Translational and Clinical Research Institute, Newcastle University, Framlington Place, Newcastle Upon Tyne NE2 4HH, UK
4 Scientific Employee with a Honorary Contract at Deutsches Zentrum für Neurodegenerative Erkrankungen (DZNE), 37075 Göttingen, Germany
* Correspondence: touteir@gwdg.de; Tel.: +49-(0)-551-391-3544; Fax: +49-(0)-551-392-2693

Abstract: The accumulation of misfolded alpha-synuclein (aSyn) throughout the brain, as Lewy pathology, is a phenomenon central to Parkinson's disease (PD) pathogenesis. The stereotypical distribution and evolution of the pathology during disease is often attributed to the cell-to-cell transmission of aSyn between interconnected brain regions. The spreading of conformationally distinct aSyn protein assemblies, commonly referred as strains, is thought to result in a variety of clinically and pathologically heterogenous diseases known as synucleinopathies. Although tremendous progress has been made in the field, the mechanisms involved in the transfer of these assemblies between interconnected neural networks and their role in driving PD progression are still unclear. Here, we present an update of the relevant discoveries supporting or challenging the prion-like spreading hypothesis. We also discuss the importance of aSyn strains in pathology progression and the various putative molecular mechanisms involved in cell-to-cell protein release. Understanding the pathways underlying aSyn propagation will contribute to determining the etiology of PD and related synucleinopathies but also assist in the development of new therapeutic strategies.

Keywords: Parkinson's disease; alpha-synuclein; prion-like spreading; cell-to-cell transfer; neurodegeneration

Citation: Brás, I.C.; Outeiro, T.F. Alpha-Synuclein: Mechanisms of Release and Pathology Progression in Synucleinopathies. *Cells* **2021**, *10*, 375. https://doi.org/10.3390/cells10020375

Academic Editor: Cristine Alves Da Costa

Received: 23 January 2021
Accepted: 9 February 2021
Published: 12 February 2021

1. Introduction

More than 200 years ago, James Parkinson described some of the clinical symptoms of the disease that, later, was named as Parkinson's disease (PD) in "Essay on the Shaking Palsy" [1]. Clinically, patients exhibit a progressive deterioration of neurological functions such as cognition and motor function, but also sleep disorders (rapid eye movement (REM) sleep disorder), hyposmia, and autonomic failure [2]. The motor symptoms result from the severe loss of dopaminergic neurons in the substantia nigra pars compacta (SNpc) and the consequent deregulation of basal ganglia activity [3]. PD is one of several synucleinopathies, which are a diverse group of neurodegenerative diseases known for the deposition of misfolded alpha-synuclein (aSyn) in the brain [4,5]. aSyn can accumulate in Lewy bodies (LBs) or Lewy neurites (LNs), in Lewy body diseases, or in glial cytoplasmic inclusions (GCIs), in multiple system atrophy (MSA) [6–8].

LBs inclusions contain high levels of aSyn phosphorylated at serine 129 (pS129). In addition, it is estimated that 10 to 30% aSyn in LBs is truncated in the N- or C-terminal regions [9–11]. In the brain, synaptic dysfunction and neuronal loss are thought to precede the formation of aSyn inclusions.

Detailed observations of aggregated aSyn in post-mortem brain tissue from patients at different clinical stages of PD form the basis for the hypothesis that Lewy pathology can progress, as disease progresses, through interconnected brain regions [12–14]. The putative neurotoxicity exerted by aSyn aggregates, and due to cell-to-cell transfer, it is correlated with increased severity of the clinical symptoms of PD [15]. Although the molecular mechanisms involved in disease progression remain unclear, several studies in human tissue and in animal models are consistent with the cell-to-cell transmission of pathological aSyn [16]. Importantly, recent evidence suggests that transfer of aSyn aggregates with disease-specific conformations, referred to as strains, may partly explain the existence of distinct synucleinopathies (Figure 1) [17–20]. Similarly to what happens in prion diseases, aSyn strains are thought to template the aggregation of native aSyn into pathological forms, resulting in the spreading and progression of disease pathology [21]. Currently, the molecular mechanisms and factors modulating aSyn aggregation remain obscure, highlighting the need for further studies.

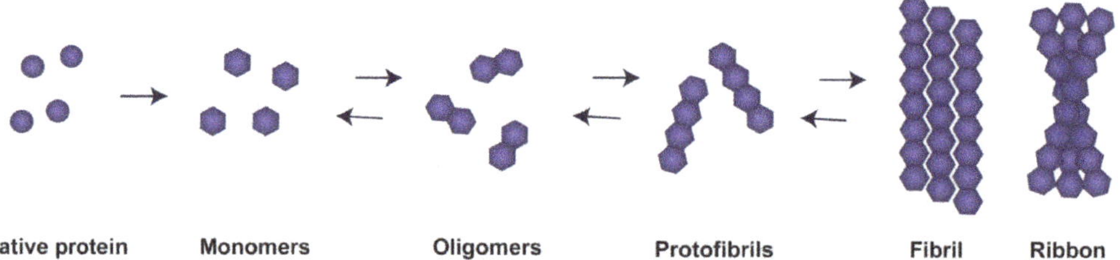

Figure 1. Model for templated misfolding of endogenous alpha-synuclein (aSyn). Under pathological conditions, due to genetic or environmental factors, natively unfolded aSyn monomers are able to self-aggregate in pathological oligomers. These species can be extended into protofibrils and other mature species such as fibrils or ribbons that deposit into inclusion bodies as Lewy bodies (LBs) and Lewy neurites (LNs). Although the biophysical properties and formation of ribbons are still not well understood, the other aSyn assemblies coexist in a dynamic equilibrium and can be transformed into higher- or lower-order conformations.

In this review, we focus on selected studies supporting the prion-like behavior of aSyn and on the molecular mechanisms involved in the spreading of pathology during disease progression. Further clarification of these concepts will assist in the development of new therapeutic strategies aimed at preventing disease progression in synucleinopathies.

2. aSyn: From Function to Neurotoxicity

The synuclein family comprises three small soluble proteins, alpha-, beta- and gamma-synuclein, that bind to phospholipid membranes [22]. aSyn is encoded by the *SNCA* gene and is composed of three distinct domains, which are defined on their amino acid composition: The N-terminal lipid-binding domain, an amyloid-binding central region (NAC), and a C-terminal disordered region [23]. The N-terminal domain is positively charged and contains seven amphipathic repeats containing a conserved KTKEGV hexameric motif, which enables an alpha-helical structure and interactions with membranes [24,25]. The central NAC region is hydrophobic and is mainly involved in fibril formation and aggregation [26]. Lastly, the C-terminal domain is highly acidic and is used for interaction with metals, small molecules, proteins, and other aSyn domains [27].

aSyn is abundant in the brain, although it also exists in erythrocytes and platelets, as well as in other tissues [28]. In the brain, under normal conditions, aSyn is mainly expressed in neuronal cells, located in the pre-synaptic terminal, and possibly bound to membranes of synaptic vesicles [29–31]. Although the precise function of aSyn is still a matter of debate, it is thought to play a role in the recycling of synaptic vesicles [32,33]. In particular, aSyn inhibits synaptic vesicle release and disrupts the SNARE complex-mediated lipid membrane fusion [34,35]. A recent study demonstrated that aSyn interacts with VAMP2 to cluster the synaptic vesicle pools, attenuating their recycling [36].

aSyn exists in equilibrium between the unfolded form in the cytosol and an alpha-helical-rich form when bound to membranes [28,37,38]. Intriguingly, under physiological conditions, aSyn was also described to exist as helically folded tetramers that might be more resistant to aggregation, but these findings remain controversial [39]. In disease conditions, it forms beta-sheet-rich amyloid fibrils that accumulate in the brains of patients [38]. In the aggregation process, natively unfolded aSyn monomers are able to self-aggregate into pathological oligomers and, subsequently, into insoluble fibrils (Figure 1). Interestingly, the interaction of aSyn with dopamine results in its oxidation and in the accumulation of aSyn protofibrils, possibly explaining the increased vulnerability of dopaminergic neurons in PD [40].

In addition to neuropathological evidence, genetic evidence associates aSyn with the pathogenesis of PD and other synucleinopathies. Point mutations in the gene encoding for aSyn, as well as genomic duplications or triplications, result in familial forms of PD [41–43]. Presently, six missense mutations in the *SNCA* gene have been associated with autosomal dominant PD (A53T, A30P, E46K, H50Q, G51D, and A53E) [44–49]. The mutations are clustered within the membrane-binding domain, suggesting the contribution of this region to aSyn dysfunction [50–52].

The specific factors that trigger aSyn aggregation still remain unclear. Mutations, expression levels, clearance efficiency, saturation of membranes, environmental factors, interactions with other amyloidogenic proteins, and/or with intermediary toxic species, truncation, or post-translational modifications are among the myriad of possible factors [23].

The pathological consequences of aSyn dyshomeostasis may themselves exacerbate such dyshomeostasis. These include dysregulation of mitochondrial activity, impairment of endoplasmic reticulum (ER)-Golgi and of synaptic vesicle trafficking, disruption of plasma membrane integrity, impairment of protein clearance systems, or impaired immune system and inflammation responses [23,53,54].

3. The Concept of aSyn Prionoids and Strains

Prions are misfolded and infectious protein assemblies that are capable of transmitting and propagating a disease [55]. Prions arise due to the aberrant folding of endogenous native cellular prion protein (PrP^{C}) into an altered form, which is known as scrapie (PrP^{Sc}) [56]. A remarkable feature of PrP^{Sc} is its ability to spread from an infected to a healthy cell, causing the self-propagation of the toxic species throughout the brain. Other characteristic properties of prions include the ability to exist with distinct stable conformations, which are commonly referred as strains. Prions are interindividual transmissible and are the cause of transmissible spongiform encephalopathies (TSE) such as Creutzfeldt–Jakob disease (CJD) or fatal familial insomnia in humans, or mad cow disease in bovines [57,58].

Over the last decade, the terms "prionoid" and "prion-like" have been used to describe the self-propagation, through seeding, of disease-related proteins in an analogy with prion disease [59]. In particular, they define the ability of misfolded proteins to recruit physiological proteins of the same type (to seed) and to induce their conversion into a pathological form that propagates from cell-to-cell. However, the use of this terminology has been one of the most controversial topics in the field, since there is currently no evidence demonstrating the direct transmission of neurodegenerative diseases between individuals, contrary to prion diseases.

Stanley Prusiner, who received the Nobel Prize for his work on prion diseases, proposed in 1987 that misfolded proteins associated with other neurodegenerative diseases might have similar properties. The propagation of these proteins would require a permissive host, a suitable environment for replication and transmission, and possibly long incubation times [59]. Recently, growing in vivo and in vitro experimental evidence has shown that templated conversion may not only be characteristic of PrP^{Sc} but also of other disease-related proteins, such as aSyn, tau, or beta-amyloid (Abeta) [57,60–63].

Strikingly, the aggregation of endogenous aSyn can occur through a homotypic (self-seeding) or heterotypic seeding [64–72]. The first term is referred to aSyn templating that requires the presence of the hydrophobic NAC region [26,73]. In contrast, heterotypic seeding refers to the involvement of other proteins in the initiation of aSyn aggregation and pathogenesis (such as Abeta, tau, or huntingtin) [74–77].

The first description of protein "strains" came from the observation of distinct clinical phenotypes in animals after infection with PrP^{Sc} [78]. In humans, at least four different PrP^{Sc} strains exist. They present distinct glycosylation patterns and lead to distinct clinical symptoms, anatomical distribution of the pathology, transmission properties, and seeding proficiencies. It is currently established that the strain-specific properties are encoded in the structure of the misfolded proteins, and these are maintained during the continuous transfer in vitro and in vivo [79]. More recently, it has been proposed that distinct structural conformations, or strains, can also be a feature of other disease-associated proteins, thereby explaining the diverse pathological and clinical phenotypes observed in different neurodegenerative diseases. In fact, it was proposed that the heterogeneity of synucleinopathies might be partly attributed to the accumulation of strains with distinct aSyn conformations. In this context, aSyn strains are assemblies that exhibit distinct biochemical, structural and physical characteristics and are thought to have different seeding and spreading capacities [17,80–82]. Interestingly, the pathological form of aSyn has a beta-sheet-rich structure similar to PrP^{Sc}. Other similarities include the abnormal folding of endogenous protein into different strains via a template protein, transfer of misfolded proteins between cells, and pathology propagation in the brain [20,83,84]. Intriguingly, the heterogeneity of synucleinopathies and other neurodegenerative diseases is not usually attributed to an alternate hypothesis that arises from the neuropathological examination of the brain: The simultaneous presence of multiple types of pathologies which could, depending on the relative levels, explain the heterogeneity of the multi-morbid old brain [85].

Nevertheless, several recent studies have uncovered apparent conformational differences in aSyn assemblies among different disorders. For instance, MSA strains have shown different seeding potencies and conformations when compared with the strains present in PD brains [19,84,86–88]. In particular, aSyn MSA seeds maintain strain characteristics following successive propagation, and they are more resistant to proteolysis and to inactivation with formalin, similar to PrP^{Sc} [89]. Consistently, MSA pathology progresses more rapidly than PD, suggesting that the pathological seeds in MSA are more toxic and spread rapidly throughout the brain [80,81,89,90].

The molecular origin for distinct aSyn strains in humans remains largely unknown. Protein conformations, post-translational modifications, local cellular milieu, or even different cell types can form the basis for distinct bioactivities that, at least in part, explain the heterogeneity of neurodegenerative diseases. In the future, it will be important to investigate how the structural characteristics of different aSyn strains can explain the diverse phenotypes in synucleinopathies.

4. Braak Staging and Prion-Like Spreading Hypothesis

In 2003, Braak and co-workers proposed a staging model for categorizing the progression of pathology in PD through neuroanatomically interconnected regions in the brain [13,14]. This model proposes that environmental factors, such as toxins or inflammatory agents, might trigger the formation of LB pathology in the enteric nervous system (ENS), particularly in the gut, or in the olfactory bulb, which would then spread into the brain [91–95], via the vagus nerve in the direction to the substantia nigra pars compacta [96]. More recently, aSyn inclusions were identified in the heart and stomach of a rat model injected with aSyn assemblies into the duodenum. This suggests an anterograde spreading of aSyn pathology (dorsal motor nucleus of the vagus nerve to the stomach), which is followed by a primary retrograde mechanism [97]. These observations indicate the susceptibility of different neuronal populations to aging, demonstrating a unique spatiotemporal distribution of pathology. Another explanation might be the intercellular transfer of unknown pathogens through preferential routes, resulting in the stereotypical progression of pathology.

In 2008, aSyn-positive LBs were observed in grafted fetal mesencephalic dopaminergic neurons that were transplanted in the striatum of PD patients in an effort to alleviate clinical symptoms [98,99]. The phenomenon of Lewy pathology in grafted neurons was interpreted as having been caused by the transfer of aggregated aSyn seeds to the healthy neurons, supporting the hypothesis that PD might be a prion-like disease [69].

5. Lewy Pathology: More Than Simply One Mechanism or Hypothesis

Several post-mortem observations support the Braak hypothesis. Lewy pathology is observed in the olfactory bulb in the majority of PD cases [100], and reduced olfaction (hyposmia) is an early indicator of PD [101–103]. However, hyposmia is not a reliable indicator in some of the genetic forms of the disease [104].

Alterations in the brain–gut–microbiota axis, as enteric pathology and gastrointestinal symptoms, have also been documented in several studies [96,105–108]. These evidences support the hypothesis that PD pathology can spread from the gut to the brain [109]. Recent epidemiological studies also indicate that truncal vagotomy or appendectomy reduces the risk of developing PD, providing support for the gut-to-brain hypothesis [107,110–112]. In animal models, the injection of aSyn assemblies in the olfactory bulb results in their spreading to the brainstem [113–115]. Interestingly, injection in the gastrointestinal tract resulted in the formation of aSyn inclusions in the brain, supporting that Lewy pathology can spread from gut to brain [93,97,116–119].

However, Braak staging does not explain all clinical cases and abnormal distribution of aSyn pathology [120]. Strikingly, elderly patients with progressive Lewy pathology can lack clinical symptoms [121,122]. In contrast, patients with advanced symptoms and certain genetic forms of PD lack widespread Lewy pathology [123–128]. Intriguingly, the presence of only peripheral pathology in the several post-mortem examinations that have been conducted during the past years was never observed, indicating that the spreading of pathology might not be a driver of disease [129,130].

Another weakness of Braak staging is the use of the selective vulnerability of neuronal types, which could make them less capable at clearing aSyn aggregates or more prone for generating aggregated species, to explain the pattern of aSyn pathology distribution. Furthermore, the pattern and propagation of aSyn pathology does not always follow neuroanatomic connectivity, suggesting that other mechanisms, besides trans-synaptic spreading, can be involved in the aSyn distribution throughout the brain [65,131,132]. In addition, no LB pathology was observed in a 14-year-old graft transplantation, indicating that the presence or not of LBs in the patient brain grafts can be associated with the graft environment, the time post-grafting, and individual differences between PD patients [133,134]. This raises the possibility that pathology might be initiated by the microenvironment of the PD brain and not through the cell-to-cell transfer of aSyn [135–137]. This would also explain why not all PD patients develop Lewy pathology in the ENS [105,129,138].

While human studies have suggested that aSyn pathology might be transmissible intra- and inter-cellularly, the exact nature of the endogenous seeds responsible for this process remains unknown [4]. aSyn oligomers, fibrils, ribbons, and pre-formed fibrils are examples of different types of recombinant strains that can be generated using different chemical/biochemical conditions and have shown distinct cell type preference and neurotoxicity (Figure 1) [139,140]. For example, different buffer and salt conditions can enable the formation of either classic fibrils or twisted assemblies as ribbons [17,139]. The effect of aSyn strains in disease propagation and the study of their propagation from host to grafted tissue has been addressed in several studies by the injection of aSyn assemblies in animal models [69,141–143]. After injection, these seeds can cross the blood–brain barrier and reach the central nervous system [144]. While fibrils injection in mice causes a loss of dopamine neurons and motor defects, ribbons result in the formation of aSyn inclusions in oligodendrocytes and replicate a pathological marker of MSA. After injection, recombinant aSyn assemblies can imprint their intrinsic structures by conversion of the endogenous monomeric protein [139,145]. Distinct aSyn strains can also be generated by consecutively passaging aSyn fibrils in cells [140].

Multiple lines of evidence suggest that oligomeric aSyn species, which are thought to precede the fibrillar aggregates found in LBs, are the culprits for seeding and neuronal degeneration in PD (Figure 1) [146,147]. The assessment of the impact of these oligomeric species in the formation of aSyn inclusions is usually difficult, because oligomers are inherently transient forms and quickly recruit monomeric aSyn to form fibrils [17,148,149]. Therefore, the term "oligomer" is broad and unspecific, constituting a source of unclarity in the field. Interestingly, it has been proposed that the oligomer concentrations that result in toxicity are different from those that efficiently seed the self-amplification [150]. A minor loss of dopaminergic neurons is observed in animal models after the injection of oligomers into the striatum. In contrast, short fibrillar fragments considerably decrease the number of dopaminergic neurons and result in the formation of aSyn inclusions in the cortex and amygdala. Remarkably, short fibril fragments show stronger effects that are attributed to their ability to recruit monomeric aSyn and spread in vivo and hence contribute to the development of PD-like phenotypes [151]. A number of key questions regarding oligomer toxicity and propagation remain to be elucidated. For instance, the role of oligomers in the cell-to-cell propagation across anatomical connected pathways, and the factors that lead to oligomer formation and accumulation. Clarifications of these topics are particularly important for the development of immunotherapy approaches aimed at targeting toxic forms of aSyn.

Interestingly, aSyn assemblies can be originated from the conversion of the endogenous protein, or through the disaggregation of amyloid fibrils by chaperones that produces both monomeric and oligomeric aSyn [152]. In particular, the chaperone HSP110 diminishes the formation of aSyn aggregates in the brain [153], suggesting a mechanism where these oligomers could seed endogenous competent oligomers that could later be propagated from cell-to-cell. Another possibility is the fragmentation of aSyn aggregates by lysosomal proteases and the release of smaller seeding-competent conformers of aSyn. In fact, low pH increases fibril fragmentation, and it might be replicated in endosomes and lysosomes due to their acidic pH [154]. Characterization of the aSyn seeds produced by disaggregases and protein degradation pathways will assist in the development of therapeutic strategies that modulate aSyn levels in the cells.

Full-length, truncated, and cleaved forms of aSyn can exist intra- and extracellularly. Recently, the relevance of aSyn fragments in the extracellular space was shown not only for spreading but also for aggregation and the formation of different strains. These fragments are also able to mediate the aggregation of endogenous full-length aSyn [155].

Pathogenic mutations can also facilitate the intercellular transfer and cytotoxicity of aSyn, contributing to early disease onset and to more rapid progression. For example, it was shown that H50Q and A53T mutations significantly increased aSyn secretion. Furthermore, H50Q, G51D, and A53T pre-formed fibrils efficiently seeded in vivo and acutely induced neuroinflammation [156]. These data indicate that pathogenic mutations augment the prion-like spread of aSyn.

Mutations in the GBA gene, which encodes the lysosomal enzyme glucocerebrosidase (GCase), are an important genetic risk factor for PD. GCase activity is also reduced in sporadic PD brains and with aging. Loss of GCase activity impairs the autophagy lysosomal pathway, resulting in increased aSyn levels. Furthermore, elevated levels of aSyn result in decreased GCase activity, suggesting that GCase deficiency increases the spreading of aSyn pathology and likely contributes to the earlier age of onset and augmented cognitive decline associated with GBA-PD [157].

6. Inconsistencies in the aSyn Cell-to-Cell Spreading Model

A wide range of studies is consistent with the prion-like spreading of aSyn. However, there are several points that still need further clarification. One of the main points against this model is the lack of studies demonstrating that aSyn can be transmitted between individuals [158,159]. In traditional prion diseases, transmission occurs between individuals of the species or even across different species. An important and obvious difference between PrP^C and aSyn is the transmembrane nature and extracellular location of PrP^C, while aSyn is predominantly intracellular.

Another hypothesis is that aSyn transfer occurs through passive release from damaged or dead neurons and not via an active mechanism. The amount of aSyn in the host cell is a key determinant of aSyn pathology generation and spreading, and it remains to be seen whether PD serves as a reservoir of aSyn in a manner similar to what is observed in wild-type or transgenic animals [160].

Additionally, there is still little evidence demonstrating that human brain-derived pathological aSyn can spread. If spreading is an important factor in the progression of PD, then evidence needs to be obtained showing the progressive spread of endogenous localized aSyn pathology through connected circuits (comparable to pre-formed seeding models).

Although there are several studies associating aSyn strains with synucleinopathy pathogenesis, the results found in the literature are not always consistent [63]. This may be due to variability in the methodologies, protocols for the preparation of aSyn assemblies [161], genetic background of the animal model, amount of exogenous aSyn assemblies injected in the animal brain, interference with the expression of mouse aSyn, and time post-injection when the samples were collected. Much greater standardization is needed in all these parameters to enable the comparison of the various results.

The ability of human aSyn seeds to induce the formation of inclusions in animal models is another source of controversy. While some studies demonstrated cross-seeding effects between human and mouse aSyn, other studies described the existence of a species barrier. The compatibility between the exogenous aSyn seeds and the endogenous protein has been suggested as a key element of seeding activity in PD models [162].

7. Mechanisms for Cell-to-Cell Transfer of aSyn

Surprisingly, as it is considered a cytosolic protein, aSyn is present in several human biofluids including saliva, plasma, cerebral spinal fluid (CSF), and red blood cells [163–167]. Major efforts are underway in an attempt to use extracellular aSyn as a biomarker in synucleinopathies, but the correlation between systemic aSyn levels with disease progression remains a matter of debate, in part due of the identification of aSyn assemblies in healthy controls [164,168–170].

Non-classic secretory pathways have been proposed to be involved in the release of aSyn from cells. These include both passive and active mechanisms (Figure 2). Passive mechanisms include diffusion through the cell membrane and release through compromised cell membranes. Monomeric aSyn, but not higher-order assemblies such as oligomers or aggregates, can diffuse through the cell membrane [171,172]. This process possibly relies on a membrane translocator, since aSyn cannot pass the lipid bilayer [172,173]. It was recently described that aSyn can also be transferred via gap junctions, which are present between adjacent cells [174,175]. Interestingly, endogenous aSyn localized in the cytoplasm remains trapped inside the cell when compared with exogenously added aSyn that can be taken up and released from the cell via diffusion [172]. This suggests that the release of aSyn through compromised cell membranes has a minor effect in the process [176].

A recent study identified 14-3-3 proteins as potential regulators of aSyn transmission, proposing that under dysfunction, they may mediate aSyn oligomerization and seeding [177]. Furthermore, the formation of other aSyn assemblies or post-translational modifications may prevent endogenous aSyn from passively escaping the cell.

A fraction of the cellular aSyn can be actively secreted via non-classical ER/Golgi-independent exocytosis. The folding state-dependent release of aSyn has been shown in several cell types and is most probably correlated with their function in the cell [70,143,173,178–180]. This process also occurs under oxidative stress [181], stress conditions [182], or dopamine treatment [183]. Interestingly, the quantity of aSyn released to the cell media is correlated with the intracellular levels [184]. In addition, the susceptibility of different neuronal populations is linked to their endogenous aSyn expression level, establishing that endogenous aSyn levels play a key role in aSyn prion-like seeding [185].

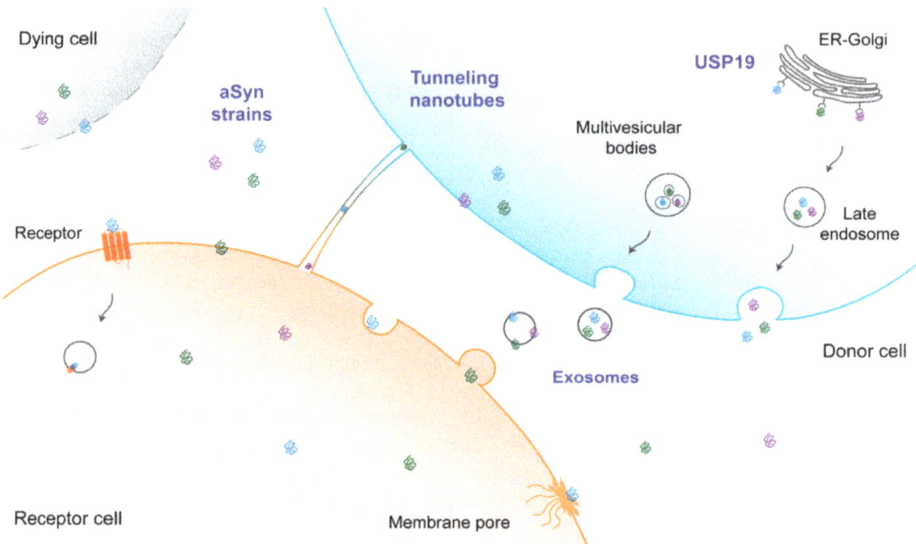

Figure 2. Schematic representation of the possible molecular mechanisms involved in the cell-to-cell transmission and progression of aSyn pathology in Parkinson's disease (PD). Release of aSyn to the extracellular space can occur via exocytosis/direct translocation through the plasma membrane from a donor to a recipient cell (blue cell). Additionally, misfolded-associated protein secretion pathway (MAPS) is also used to preferentially export aberrant cytosolic proteins. In this mechanism, the endoplasmic reticulum (ER)-associated deubiquitylase USP19 recruits misfolded proteins to the ER surface for deubiquitylation. Then, these cargoes are encapsulated into ER-associated late endosomes and secreted to the extracellular space. Exosomes are derived from multivesicular bodies (MVBs) and have been reported to mediate aSyn release from cells. Tunneling nanotubes (TNTs) can form a direct connection between two cells possibly allowing aSyn from one cell to another. The entry of aSyn into the receptor cell can occur via passive diffusion through the plasma membrane, endocytosis, receptor-mediated endocytosis, and exosome-mediated transfer (orange cell). Furthermore, a high concentration of aSyn in the membrane potentiates its oligomerization and the putative formation of trans-membrane amyloid pores (these pores have yet to be identified in human brain tissue). Last, dying neurons will release their content into the extracellular space, which is another potential source of extracellular aSyn (gray cell).

More recently, the misfolding-associated protein secretion pathway (MAPS) has been identified as an unconventional secretion pathway through an ER-dependent process, for preferentially exporting aberrant cytosolic proteins, including aSyn (Figure 2) [186]. The ER-associated deubiquitylase USP19 contains a catalytic domain with a chaperone activity that allows the recruitment of misfolded proteins to the ER surface for deubiquitylation Then, the deubiquitylated proteins are encapsulated into late endosomes and secreted to the extracellular space [186]. In addition, the HSP70 co-chaperone DNAJC5 was described to play a key role in the secretion of aggregated aSyn assemblies MAPS [187].

Another mechanism involved in the release of aSyn assemblies is through their association with exosomes (Figure 2). Exosomes are small vesicles produced from the fusion of multivesicular bodies (MVBs) with the plasma membrane, resulting in their release to the extracellular space [188]. Exosomes are secreted from various cell types, including neurons, astrocytes, and microglia, and they have a regulatory function in synapses and in the intercellular exchange of membrane proteins [189]. Interestingly, vesicular aSyn is more prone to aggregation than cytoplasmic aSyn, and exosomes isolated from the CSF of patients exhibit higher seeding potency compared with controls [173,190]. It has been described that aSyn can be released by exosomes in a calcium-dependent manner

This can be further exacerbated after lysosomal inhibition, lipid peroxidation, or ATG5 knockdown [173,191–195]. Stress conditions increase the translocation of aSyn into vesicles, thus causing its subsequent release to the extracellular space [182]. Furthermore, exosomes released from microglial cells can play an active role in the process of aSyn transmission to neurons. This process is further enhanced by the release of pro-inflammatory cytokines, resulting in protein aggregation and spreading [196].

Tunneling nanotubes (TNTs) represent a novel type of intercellular communication machinery (Figure 2). These membranous structures mediate the communication between neighboring cells and have been implicated in the transfer of pathological aSyn aggregates [178]. However, it remains unclear whether these structures actually mediate the connection of different cell types in vivo.

Passive diffusion [171], conventional endocytosis [69,142,172,197,198], direct penetration through the plasma membrane [199–201], and receptor-mediated endocytosis [202–205] have been proposed as pathways involved in the internalization of aSyn (Figure 2).

The reasons for aSyn secretion still remain an open question in the context of the prion-like spreading hypothesis. Technical questions relate to the low levels of aSyn and extracellular vesicles that are secreted by cultured cells to the media, and the detection of different types of aSyn species that might be present in the media. The identification of the molecular mechanisms and proteins responsible for the recognition and secretion of aSyn assemblies may, ultimately, support the development of novel approaches to prevent disease progression, but such research is only in its infancy.

8. Conclusions

Emerging evidence supports the concept that cell-to-cell transmission and disease-selective strains underlie disease progression and heterogeneity in synucleinopathies. aSyn propagation is coincident with the progression of PD pathology throughout the brain. Our current understanding of this phenomenon is that aSyn pathology spreading may not be the main driving factor in PD. However, it is not yet well understood how aggregated aSyn can transfer from cell-to-cell to induce synaptotoxicity and neurodegeneration. Additionally, it should be determined if endogenous aSyn pathology spreads between cells. Additional studies of the aggregation process are necessary in order to understand the precise mechanism of aSyn propagation in the brain.

Although several studies use in vivo models to address the prion-like properties of exogenous aSyn strains, self-propagation of the endogenous protein remains to be shown. Furthermore, the different mechanisms described to be involved in aSyn cell-to-cell spreading were observed using in vitro models, but whether these mechanisms occur in the brain of PD patients is still unknown. Overall, several studies suggest that aSyn seeds can be transferred through various cell types, inducing the aggregation of the endogenous protein.

Regardless, the different factors promoting progressive cell-to-cell transfer of aSyn need to be investigated. In the future, it will be important to understand the exact contribution of different aSyn species to the prion-like spreading of PD pathology, by elucidating their transfer and seeding properties, as well as their toxic effects on recipient cells. Clarification of these questions might support the development of novel types of interventions for PD.

Moreover, it remains unclear whether and how familial PD-associated aSyn mutants propagate throughout the brain, as they present distinct aggregation kinetics and different physicochemical properties. Since most studies do not use patient-derived aSyn assemblies, it is uncertain the extent to which experimental models recapitulate the cell-to-cell transmission in PD.

The spreading of aSyn pathology (spatiotemporal distribution, affected cell types, and morphology) in the nervous system is defined by several factors. These factors should differently influence the spread of pathology among strains, thereby causing distinct disease entities. Therefore, it may be necessary to use disease-specific aggregates in experiments in order to identify therapeutic targets that may be unique among these diseases. In prion disorders, approaches targeting PrP^C oligomers are being developed after the observation that only oligomers, not monomers, are infectious. However, considering the limited availability of human brain material, it is indispensable to develop new methodologies that enable the production of sufficient amounts of disease-specific aggregates for research.

The development of new therapeutic strategies has been slow and difficult due to the plethora of possible targets that may be tackled in synucleinopathies. This includes aSyn production, aggregation, toxicity, degradation, and spreading. The use of receptor blocking strategies to inhibit aSyn internalization, or of strain-specific antibodies to decrease the levels of extracellular aSyn, may delay the spreading of pathology, but this also needs to be investigated further.

In total, a deeper understanding of the molecular mechanisms underlying aSyn aggregation and intercellular propagation is important for understanding the pathogenesis of PD and related synucleinopathies, to identify new disease targets, and to develop novel therapeutic strategies to halt disease progression.

Author Contributions: I.C.B. and T.F.O. wrote the manuscript. All authors have read and agreed to the published version of the manuscript.

Funding: This research was funded by the Deutsche Forschungsgemeinschaft (DFG, German Research Foundation) under Germany's Excellence Strategy—EXC 2067/1- 390729940), and European Union's Horizon 2020 research and innovation program under grant agreement No. 721802 (SynDegen).

Institutional Review Board Statement: Not applicable.

Informed Consent Statement: Not applicable.

Data Availability Statement: Data sharing not applicable.

Conflicts of Interest: The authors declare no conflict of interest.

Abbreviations

aSyn	Alpha-synuclein
NAC	Amyloid-binding region
Abeta	Beta-amyloid
CSF	Cerebral spinal fluid
CJD	Creutzfeldt-Jakob disease
DLB	Dementia with Lewy bodies
ER	Endoplasmic reticulum
ENS	Enteric nervous system
GCIs	Glial cytoplasmic inclusions
GCase	Glucocerebrosidase
LBs	Lewy bodies
LNs	Lewy neurites
MAPS	Misfolded-associated protein secretion
MVBs	Multivesicular bodies
MSA	Multiple system atrophy
PD	Parkinson's disease
pS129	Phosphorylation at serine 129 in aSyn
PrP^C	Prion protein
PAF	Pure autonomic failure
REM	Rapid eye movement
PrP^{Sc}	Scrapie prion protein
SNpc	Substantia nigra pars compacta
TSE	Transmissible spongiform encephalopathies
TNTs	Tunneling nanotubes

References

1. Parkinson, J. An essay on the shaking palsy. 1817. *J. Neuropsychiatry Clin. Neurosci.* **2002**, *14*, 223–236. [CrossRef]
2. Takamatsu, Y.; Fujita, M.; Ho, G.J.; Wada, R.; Sugama, S.; Takenouchi, T.; Waragai, M.; Masliah, E.; Hashimoto, M. Motor and Nonmotor Symptoms of Parkinson's Disease: Antagonistic Pleiotropy Phenomena Derived from alpha-Synuclein Evolvability? *Parkinsons Dis.* **2018**, *2018*, 5789424. [CrossRef] [PubMed]
3. Dauer, W.; Przedborski, S. Parkinson's disease: Mechanisms and models. *Neuron* **2003**, *39*, 889–909. [CrossRef]
4. Halliday, G.M.; Holton, J.L.; Revesz, T.; Dickson, D.W. Neuropathology underlying clinical variability in patients with synucleinopathies. *Acta Neuropathol.* **2011**, *122*, 187–204. [CrossRef] [PubMed]
5. McCann, H.; Stevens, C.H.; Cartwright, H.; Halliday, G.M. Alpha-Synucleinopathy phenotypes. *Parkinsonism Relat. Disord.* **2014**, *20 Suppl 1*, S62–S67. [CrossRef]
6. Poewe, W.; Seppi, K.; Tanner, C.M.; Halliday, G.M.; Brundin, P.; Volkmann, J.; Schrag, A.E.; Lang, A.E. Parkinson disease. *Nat. Rev. Dis. Primers* **2017**, *3*, 17013. [CrossRef] [PubMed]
7. Spillantini, M.G. Parkinson's disease, dementia with Lewy bodies and multiple system atrophy are alpha-synucleinopathies. *Parkinsonism Relat. Disord.* **1999**, *5*, 157–162. [CrossRef]
8. Spillantini, M.G.; Crowther, R.A.; Jakes, R.; Cairns, N.J.; Lantos, P.L.; Goedert, M. Filamentous alpha-synuclein inclusions link multiple system atrophy with Parkinson's disease and dementia with Lewy bodies. *Neurosci. Lett.* **1998**, *251*, 205–208. [CrossRef]
9. Fujiwara, H.; Hasegawa, M.; Dohmae, N.; Kawashima, A.; Masliah, E.; Goldberg, M.S.; Shen, J.; Takio, K.; Iwatsubo, T. Alpha-Synuclein is phosphorylated in synucleinopathy lesions. *Nat. Cell Biol.* **2002**, *4*, 160–164. [CrossRef]
10. Saito, Y.; Kawashima, A.; Ruberu, N.N.; Fujiwara, H.; Koyama, S.; Sawabe, M.; Arai, T.; Nagura, H.; Yamanouchi, H.; Hasegawa, M.; et al. Accumulation of phosphorylated alpha-synuclein in aging human brain. *J. Neuropathol. Exp. Neurol.* **2003**, *62*, 644–654. [CrossRef] [PubMed]
11. Kellie, J.F.; Higgs, R.E.; Ryder, J.W.; Major, A.; Beach, T.G.; Adler, C.H.; Merchant, K.; Knierman, M.D. Quantitative measurement of intact alpha-synuclein proteoforms from post-mortem control and Parkinson's disease brain tissue by intact protein mass spectrometry. *Sci. Rep.* **2014**, *4*, 5797. [CrossRef] [PubMed]
12. Braak, H.; Del Tredici, K.; Bratzke, H.; Hamm-Clement, J.; Sandmann-Keil, D.; Rub, U. Staging of the intracerebral inclusion body pathology associated with idiopathic Parkinson's disease (preclinical and clinical stages). *J. Neurol.* **2002**, *249* (Suppl. 3), III/1–III/5. [CrossRef]
13. Braak, H.; Del Tredici, K.; Rub, U.; de Vos, R.A.; Jansen Steur, E.N.; Braak, E. Staging of brain pathology related to sporadic Parkinson's disease. *Neurobiol. Aging* **2003**, *24*, 197–211. [CrossRef]
14. Braak, H.; Rub, U.; Gai, W.P.; Del Tredici, K. Idiopathic Parkinson's disease: Possible routes by which vulnerable neuronal types may be subject to neuroinvasion by an unknown pathogen. *J. Neural Transm. (Vienna)* **2003**, *110*, 517–536. [CrossRef] [PubMed]
15. Brettschneider, J.; Del Tredici, K.; Lee, V.M.; Trojanowski, J.Q. Spreading of pathology in neurodegenerative diseases: A focus on human studies. *Nat. Rev. Neurosci.* **2015**, *16*, 109–120. [CrossRef] [PubMed]
16. Vargas, J.Y.; Grudina, C.; Zurzolo, C. The prion-like spreading of alpha-synuclein: From in vitro to in vivo models of Parkinson's disease. *Ageing Res. Rev.* **2019**, *50*, 89–101. [CrossRef] [PubMed]
17. Bousset, L.; Pieri, L.; Ruiz-Arlandis, G.; Gath, J.; Jensen, P.H.; Habenstein, B.; Madiona, K.; Olieric, V.; Bockmann, A.; Meier, B.H.; et al. Structural and functional characterization of two alpha-synuclein strains. *Nat Commun* **2013**, *4*, 2575. [CrossRef] [PubMed]
18. Pieri, L.; Madiona, K.; Melki, R. Structural and functional properties of prefibrillar alpha-synuclein oligomers. *Sci. Rep.* **2016**, *6*, 24526. [CrossRef] [PubMed]
19. Peng, C.; Gathagan, R.J.; Covell, D.J.; Medellin, C.; Stieber, A.; Robinson, J.L.; Zhang, B.; Pitkin, R.M.; Olufemi, M.F.; Luk, K.C.; et al. Cellular milieu imparts distinct pathological alpha-synuclein strains in alpha-synucleinopathies. *Nature* **2018**, *557*, 558–563. [CrossRef]
20. Peelaerts, W.; Bousset, L.; Baekelandt, V.; Melki, R. a-Synuclein strains and seeding in Parkinson's disease, incidental Lewy body disease, dementia with Lewy bodies and multiple system atrophy: Similarities and differences. *Cell Tissue Res.* **2018**, *373*, 195–212. [CrossRef]
21. Makin, S. Pathology: The prion principle. *Nature* **2016**, *538*, S13–S16. [CrossRef] [PubMed]
22. Clayton, D.F.; George, J.M. The synucleins: A family of proteins involved in synaptic function, plasticity, neurodegeneration and disease. *Trends Neurosci.* **1998**, *21*, 249–254. [CrossRef]
23. Bras, I.C.; Xylaki, M.; Outeiro, T.F. Mechanisms of alpha-synuclein toxicity: An update and outlook. *Prog. Brain Res.* **2020**, *252*, 91–129. [CrossRef]
24. Chandra, S.; Chen, X.; Rizo, J.; Jahn, R.; Sudhof, T.C. A broken alpha -helix in folded alpha-Synuclein. *J. Biol. Chem.* **2003**, *278*, 15313–15318. [CrossRef] [PubMed]
25. Ulmer, T.S.; Bax, A.; Cole, N.B.; Nussbaum, R.L. Structure and dynamics of micelle-bound human alpha-synuclein. *J. Biol. Chem.* **2005**, *280*, 9595–9603. [CrossRef] [PubMed]
26. Giasson, B.I.; Murray, I.V.; Trojanowski, J.Q.; Lee, V.M. A hydrophobic stretch of 12 amino acid residues in the middle of alpha-synuclein is essential for filament assembly. *J. Biol. Chem.* **2001**, *276*, 2380–2386. [CrossRef] [PubMed]
27. Kim, T.D.; Paik, S.R.; Yang, C.H. Structural and functional implications of C-terminal regions of alpha-synuclein. *Biochemistry* **2002**, *41*, 13782–13790. [CrossRef] [PubMed]

28. Del Tredici, K.; Braak, H. Review: Sporadic Parkinson's disease: Development and distribution of alpha-synuclein pathology. *Neuropathol. Appl. Neurobiol.* **2016**, *42*, 33–50. [CrossRef] [PubMed]
29. Burre, J. The Synaptic Function of alpha-Synuclein. *J. Parkinsons Dis.* **2015**, *5*, 699–713. [CrossRef] [PubMed]
30. Burre, J.; Sharma, M.; Sudhof, T.C. Systematic mutagenesis of alpha-synuclein reveals distinct sequence requirements for physiological and pathological activities. *J. Neurosci.* **2012**, *32*, 15227–15242. [CrossRef]
31. Logan, T.; Bendor, J.; Toupin, C.; Thorn, K.; Edwards, R.H. Alpha-Synuclein promotes dilation of the exocytotic fusion pore. *Nat Neurosci.* **2017**, *20*, 681–689. [CrossRef] [PubMed]
32. Burre, J.; Sharma, M.; Tsetsenis, T.; Buchman, V.; Etherton, M.R.; Sudhof, T.C. Alpha-synuclein promotes SNARE-complex assembly in vivo and in vitro. *Science* **2010**, *329*, 1663–1667. [CrossRef]
33. Burre, J.; Sharma, M.; Sudhof, T.C. Definition of a molecular pathway mediating alpha-synuclein neurotoxicity. *J. Neurosci.* **2015**, *35*, 5221–5232. [CrossRef] [PubMed]
34. DeWitt, D.C.; Rhoades, E. Alpha-Synuclein can inhibit SNARE-mediated vesicle fusion through direct interactions with lipid bilayers. *Biochemistry* **2013**, *52*, 2385–2387. [CrossRef]
35. Abeliovich, A.; Schmitz, Y.; Farinas, I.; Choi-Lundberg, D.; Ho, W.H.; Castillo, P.E.; Shinsky, N.; Verdugo, J.M.; Armanini, M.; Ryan, A.; et al. Mice lacking alpha-synuclein display functional deficits in the nigrostriatal dopamine system. *Neuron* **2000**, *25*, 239–252. [CrossRef]
36. Sun, J.; Wang, L.; Bao, H.; Premi, S.; Das, U.; Chapman, E.R.; Roy, S. Functional cooperation of alpha-synuclein and VAMP2 in synaptic vesicle recycling. *Proc. Natl. Acad. Sci. USA* **2019**, *116*, 11113–11115. [CrossRef]
37. Longhena, F.; Faustini, G.; Missale, C.; Pizzi, M.; Spano, P.; Bellucci, A. The Contribution of alpha-Synuclein Spreading to Parkinson's Disease Synaptopathy. *Neural Plast.* **2017**, *2017*, 5012129. [CrossRef] [PubMed]
38. Burre, J.; Sharma, M.; Sudhof, T.C. Alpha-Synuclein assembles into higher-order multimers upon membrane binding to promote SNARE complex formation. *Proc. Natl. Acad. Sci. USA* **2014**, *111*, E4274–E4283. [CrossRef] [PubMed]
39. Bartels, T.; Choi, J.G.; Selkoe, D.J. Alpha-Synuclein occurs physiologically as a helically folded tetramer that resists aggregation. *Nature* **2011**, *477*, 107–110. [CrossRef] [PubMed]
40. Conway, K.A.; Rochet, J.C.; Bieganski, R.M.; Lansbury, P.T., Jr. Kinetic stabilization of the alpha-synuclein protofibril by a dopamine-alpha-synuclein adduct. *Science* **2001**, *294*, 1346–1349. [CrossRef] [PubMed]
41. Wong, Y.C.; Krainc, D. Alpha-synuclein toxicity in neurodegeneration: Mechanism and therapeutic strategies. *Nat. Med.* **2017**, *23*, 1–13. [CrossRef]
42. Singleton, A.B.; Farrer, M.; Johnson, J.; Singleton, A.; Hague, S.; Kachergus, J.; Hulihan, M.; Peuralinna, T.; Dutra, A.; Nussbaum, R.; et al. Alpha-Synuclein locus triplication causes Parkinson's disease. *Science* **2003**, *302*, 841. [CrossRef]
43. Chartier-Harlin, M.C.; Kachergus, J.; Roumier, C.; Mouroux, V.; Douay, X.; Lincoln, S.; Levecque, C.; Larvor, L.; Andrieux, J.; Hulihan, M.; et al. Alpha-synuclein locus duplication as a cause of familial Parkinson's disease. *Lancet* **2004**, *364*, 1167–1169. [CrossRef]
44. Polymeropoulos, M.H.; Lavedan, C.; Leroy, E.; Ide, S.E.; Dehejia, A.; Dutra, A.; Pike, B.; Root, H.; Rubenstein, J.; Boyer, R.; et al. Mutation in the alpha-synuclein gene identified in families with Parkinson's disease. *Science* **1997**, *276*, 2045–2047. [CrossRef] [PubMed]
45. Kruger, R.; Kuhn, W.; Muller, T.; Woitalla, D.; Graeber, M.; Kosel, S.; Przuntek, H.; Epplen, J.T.; Schols, L.; Riess, O. Ala30Pro mutation in the gene encoding alpha-synuclein in Parkinson's disease. *Nat. Genet.* **1998**, *18*, 106–108. [CrossRef] [PubMed]
46. Zarranz, J.J.; Alegre, J.; Gomez-Esteban, J.C.; Lezcano, E.; Ros, R.; Ampuero, I.; Vidal, L.; Hoenicka, J.; Rodriguez, O.; Atares, B.; et al. The new mutation, E46K, of alpha-synuclein causes Parkinson and Lewy body dementia. *Ann. Neurol* **2004**, *55*, 164–173. [CrossRef] [PubMed]
47. Kiely, A.P.; Asi, Y.T.; Kara, E.; Limousin, P.; Ling, H.; Lewis, P.; Proukakis, C.; Quinn, N.; Lees, A.J.; Hardy, J.; et al. Alpha-Synucleinopathy associated with G51D SNCA mutation: A link between Parkinson's disease and multiple system atrophy? *Acta Neuropathol.* **2013**, *125*, 753–769. [CrossRef] [PubMed]
48. Proukakis, C.; Dudzik, C.G.; Brier, T.; MacKay, D.S.; Cooper, J.M.; Millhauser, G.L.; Houlden, H.; Schapira, A.H. A novel alpha-synuclein missense mutation in Parkinson disease. *Neurology* **2013**, *80*, 1062–1064. [CrossRef] [PubMed]
49. Pasanen, P.; Myllykangas, L.; Siitonen, M.; Raunio, A.; Kaakkola, S.; Lyytinen, J.; Tienari, P.J.; Poyhonen, M.; Paetau, A. Novel alpha-synuclein mutation A53E associated with atypical multiple system atrophy and Parkinson's disease-type pathology. *Neurobiol. Aging* **2014**, *35*, 2180.e1–2180.e5. [CrossRef] [PubMed]
50. Jo, E.; Fuller, N.; Rand, R.P.; St George-Hyslop, P.; Fraser, P.E. Defective membrane interactions of familial Parkinson's disease mutant A30P alpha-synuclein. *J. Mol. Biol.* **2002**, *315*, 799–807. [CrossRef] [PubMed]
51. Fares, M.B.; Ait-Bouziad, N.; Dikiy, I.; Mbefo, M.K.; Jovicic, A.; Kiely, A.; Holton, J.L.; Lee, S.J.; Gitler, A.D.; Eliezer, D.; et al. The novel Parkinson's disease linked mutation G51D attenuates in vitro aggregation and membrane binding of alpha-synuclein, and enhances its secretion and nuclear localization in cells. *Hum. Mol. Genet.* **2014**, *23*, 4491–4509. [CrossRef] [PubMed]
52. Ghosh, D.; Sahay, S.; Ranjan, P.; Salot, S.; Mohite, G.M.; Singh, P.K.; Dwivedi, S.; Carvalho, E.; Banerjee, R.; Kumar, A.; et al. The newly discovered Parkinson's disease associated Finnish mutation (A53E) attenuates alpha-synuclein aggregation and membrane binding. *Biochemistry* **2014**, *53*, 6419–6421. [CrossRef] [PubMed]
53. Cuervo, A.M.; Stefanis, L.; Fredenburg, R.; Lansbury, P.T.; Sulzer, D. Impaired degradation of mutant alpha-synuclein by chaperone-mediated autophagy. *Science* **2004**, *305*, 1292–1295. [CrossRef] [PubMed]

54. Emmanouilidou, E.; Stefanis, L.; Vekrellis, K. Cell-produced alpha-synuclein oligomers are targeted to, and impair, the 26S proteasome. *Neurobiol. Aging* **2010**, *31*, 953–968. [CrossRef]
55. Prusiner, S.B. Novel proteinaceous infectious particles cause scrapie. *Science* **1982**, *216*, 136–144. [CrossRef] [PubMed]
56. Prusiner, S.B. Molecular biology of prion diseases. *Science* **1991**, *252*, 1515–1522. [CrossRef]
57. Aguzzi, A.; Lakkaraju, A.K.K. Cell Biology of Prions and Prionoids: A Status Report. *Trends Cell Biol.* **2016**, *26*, 40–51. [CrossRef] [PubMed]
58. Scheckel, C.; Aguzzi, A. Prions, prionoids and protein misfolding disorders. *Nat. Rev. Genet.* **2018**, *19*, 405–418. [CrossRef] [PubMed]
59. Prusiner, S.B. Prions and neurodegenerative diseases. *N. Engl. J. Med.* **1987**, *317*, 1571–1581. [CrossRef] [PubMed]
60. Jucker, M.; Walker, L.C. Propagation and spread of pathogenic protein assemblies in neurodegenerative diseases. *Nat. Neurosci.* **2018**, *21*, 1341–1349. [CrossRef]
61. Tamguney, G.; Korczyn, A.D. A critical review of the prion hypothesis of human synucleinopathies. *Cell Tissue Res.* **2018**, *373*, 213–220. [CrossRef] [PubMed]
62. Visanji, N.P.; Brooks, P.L.; Hazrati, L.N.; Lang, A.E. The prion hypothesis in Parkinson's disease: Braak to the future. *Acta Neuropathol. Commun.* **2013**, *1*, 2. [CrossRef] [PubMed]
63. Steiner, J.A.; Quansah, E.; Brundin, P. The concept of alpha-synuclein as a prion-like protein: Ten years after. *Cell Tissue Res.* **2018**, *373*, 161–173. [CrossRef] [PubMed]
64. Volpicelli-Daley, L.A.; Luk, K.C.; Patel, T.P.; Tanik, S.A.; Riddle, D.M.; Stieber, A.; Meaney, D.F.; Trojanowski, J.Q.; Lee, V.M. Exogenous alpha-synuclein fibrils induce Lewy body pathology leading to synaptic dysfunction and neuron death. *Neuron* **2011**, *72*, 57–71. [CrossRef] [PubMed]
65. Luk, K.C.; Kehm, V.M.; Zhang, B.; O'Brien, P.; Trojanowski, J.Q.; Lee, V.M. Intracerebral inoculation of pathological alpha-synuclein initiates a rapidly progressive neurodegenerative alpha-synucleinopathy in mice. *J. Exp. Med.* **2012**, *209*, 975–986. [CrossRef]
66. Volpicelli-Daley, L.A.; Gamble, K.L.; Schultheiss, C.E.; Riddle, D.M.; West, A.B.; Lee, V.M. Formation of alpha-synuclein Lewy neurite-like aggregates in axons impedes the transport of distinct endosomes. *Mol. Biol. Cell* **2014**, *25*, 4010–4023. [CrossRef]
67. Chu, Y.; Muller, S.; Tavares, A.; Barret, O.; Alagille, D.; Seibyl, J.; Tamagnan, G.; Marek, K.; Luk, K.C.; Trojanowski, J.Q.; et al. Intrastriatal alpha-synuclein fibrils in monkeys: Spreading, imaging and neuropathological changes. *Brain* **2019**, *142*, 3565–3579. [CrossRef]
68. Luk, K.C.; Song, C.; O'Brien, P.; Stieber, A.; Branch, J.R.; Brunden, K.R.; Trojanowski, J.Q.; Lee, V.M. Exogenous alpha-synuclein fibrils seed the formation of Lewy body-like intracellular inclusions in cultured cells. *Proc. Natl. Acad. Sci. USA* **2009**, *106*, 20051–20056. [CrossRef]
69. Desplats, P.; Lee, H.J.; Bae, E.J.; Patrick, C.; Rockenstein, E.; Crews, L.; Spencer, B.; Masliah, E.; Lee, S.J. Inclusion formation and neuronal cell death through neuron-to-neuron transmission of alpha-synuclein. *Proc. Natl. Acad. Sci. USA* **2009**, *106*, 13010–13015. [CrossRef] [PubMed]
70. Freundt, E.C.; Maynard, N.; Clancy, E.K.; Roy, S.; Bousset, L.; Sourigues, Y.; Covert, M.; Melki, R.; Kirkegaard, K.; Brahic, M. Neuron-to-neuron transmission of alpha-synuclein fibrils through axonal transport. *Ann. Neurol.* **2012**, *72*, 517–524. [CrossRef] [PubMed]
71. Brahic, M.; Bousset, L.; Bieri, G.; Melki, R.; Gitler, A.D. Axonal transport and secretion of fibrillar forms of alpha-synuclein, Abeta42 peptide and HTTExon 1. *Acta Neuropathol.* **2016**, *131*, 539–548. [CrossRef]
72. Recasens, A.; Dehay, B.; Bove, J.; Carballo-Carbajal, I.; Dovero, S.; Perez-Villalba, A.; Fernagut, P.O.; Blesa, J.; Parent, A.; Perier, C.; et al. Lewy body extracts from Parkinson disease brains trigger alpha-synuclein pathology and neurodegeneration in mice and monkeys. *Ann. Neurol.* **2014**, *75*, 351–362. [CrossRef] [PubMed]
73. El-Agnaf, O.M.; Jakes, R.; Curran, M.D.; Middleton, D.; Ingenito, R.; Bianchi, E.; Pessi, A.; Neill, D.; Wallace, A. Aggregates from mutant and wild-type alpha-synuclein proteins and NAC peptide induce apoptotic cell death in human neuroblastoma cells by formation of beta-sheet and amyloid-like filaments. *FEBS Lett.* **1998**, *440*, 71–75. [CrossRef]
74. Pletnikova, O.; West, N.; Lee, M.K.; Rudow, G.L.; Skolasky, R.L.; Dawson, T.M.; Marsh, L.; Troncoso, J.C. Abeta deposition is associated with enhanced cortical alpha-synuclein lesions in Lewy body diseases. *Neurobiol. Aging* **2005**, *26*, 1183–1192. [CrossRef]
75. Tomas-Zapico, C.; Diez-Zaera, M.; Ferrer, I.; Gomez-Ramos, P.; Moran, M.A.; Miras-Portugal, M.T.; Diaz-Hernandez, M.; Lucas, J.J. Alpha-Synuclein accumulates in huntingtin inclusions but forms independent filaments and its deficiency attenuates early phenotype in a mouse model of Huntington's disease. *Hum. Mol. Genet.* **2012**, *21*, 495–510. [CrossRef]
76. Bassil, F.; Brown, H.J.; Pattabhiraman, S.; Iwasyk, J.E.; Maghames, C.M.; Meymand, E.S.; Cox, T.O.; Riddle, D.M.; Zhang, B.; Trojanowski, J.Q.; et al. Amyloid-Beta (Abeta) Plaques Promote Seeding and Spreading of Alpha-Synuclein and Tau in a Mouse Model of Lewy Body Disorders with Abeta Pathology. *Neuron* **2020**, *105*, 260–275. [CrossRef] [PubMed]
77. Badiola, N.; de Oliveira, R.M.; Herrera, F.; Guardia-Laguarta, C.; Goncalves, S.A.; Pera, M.; Suarez-Calvet, M.; Clarimon, J.; Outeiro, T.F.; Lleo, A. Tau enhances alpha-synuclein aggregation and toxicity in cellular models of synucleinopathy. *PLoS ONE* **2011**, *6*, e26609. [CrossRef] [PubMed]
78. Pattison, I.H.; Millson, G.C. Scrapie produced experimentally in goats with special reference to the clinical syndrome. *J. Comp. Pathol.* **1961**, *71*, 101–109. [CrossRef]
79. Collinge, J.; Clarke, A.R. A general model of prion strains and their pathogenicity. *Science* **2007**, *318*, 930–936. [CrossRef] [PubMed]
80. Peng, C.; Gathagan, R.J.; Lee, V.M. Distinct alpha-Synuclein strains and implications for heterogeneity among alpha-Synucleinopathies. *Neurobiol. Dis.* **2018**, *109*, 209–218. [CrossRef]

81. Melki, R. Role of Different Alpha-Synuclein Strains in Synucleinopathies, Similarities with other Neurodegenerative Diseases. *J. Parkinsons Dis.* **2015**, *5*, 217–227. [CrossRef]
82. Guerrero-Ferreira, R.; Kovacik, L.; Ni, D.; Stahlberg, H. New insights on the structure of alpha-synuclein fibrils using cryo-electron microscopy. *Curr. Opin. Neurobiol.* **2020**, *61*, 89–95. [CrossRef] [PubMed]
83. Uversky, V.N. A protein-chameleon: Conformational plasticity of alpha-synuclein, a disordered protein involved in neurodegenerative disorders. *J. Biomol. Struct. Dyn.* **2003**, *21*, 211–234. [CrossRef] [PubMed]
84. Prusiner, S.B.; Woerman, A.L.; Mordes, D.A.; Watts, J.C.; Rampersaud, R.; Berry, D.B.; Patel, S.; Oehler, A.; Lowe, J.K.; Kravitz, S.N.; et al. Evidence for alpha-synuclein prions causing multiple system atrophy in humans with parkinsonism. *Proc. Natl. Acad. Sci. USA* **2015**, *112*, E5308–E5317. [CrossRef]
85. Attems, J. The multi-morbid old brain. *Acta Neuropathol.* **2017**, *134*, 169–170. [CrossRef] [PubMed]
86. Lau, A.; So, R.W.L.; Lau, H.H.C.; Sang, J.C.; Ruiz-Riquelme, A.; Fleck, S.C.; Stuart, E.; Menon, S.; Visanji, N.P.; Meisl, G.; et al. Alpha-Synuclein strains target distinct brain regions and cell types. *Nat. Neurosci.* **2020**, *23*, 21–31. [CrossRef] [PubMed]
87. Shahnawaz, M.; Mukherjee, A.; Pritzkow, S.; Mendez, N.; Rabadia, P.; Liu, X.; Hu, B.; Schmeichel, A.; Singer, W.; Wu, G.; et al. Discriminating alpha-synuclein strains in Parkinson's disease and multiple system atrophy. *Nature* **2020**, *578*, 273–277. [CrossRef] [PubMed]
88. Woerman, A.L.; Kazmi, S.A.; Patel, S.; Freyman, Y.; Oehler, A.; Aoyagi, A.; Mordes, D.A.; Halliday, G.M.; Middleton, L.T.; Gentleman, S.M.; et al. MSA prions exhibit remarkable stability and resistance to inactivation. *Acta Neuropathol.* **2018**, *135*, 49–63. [CrossRef]
89. Woerman, A.L.; Oehler, A.; Kazmi, S.A.; Lee, J.; Halliday, G.M.; Middleton, L.T.; Gentleman, S.M.; Mordes, D.A.; Spina, S.; Grinberg, L.T.; et al. Multiple system atrophy prions retain strain specificity after serial propagation in two different Tg(SNCA*A53T) mouse lines. *Acta Neuropathol.* **2019**, *137*, 437–454. [CrossRef]
90. Yamasaki, T.R.; Holmes, B.B.; Furman, J.L.; Dhavale, D.D.; Su, B.W.; Song, E.S.; Cairns, N.J.; Kotzbauer, P.T.; Diamond, M.I. Parkinson's disease and multiple system atrophy have distinct alpha-synuclein seed characteristics. *J. Biol. Chem.* **2019**, *294*, 1045–1058. [CrossRef] [PubMed]
91. Hawkes, C.H.; Del Tredici, K.; Braak, H. Parkinson's disease: The dual hit theory revisited. *Ann. N. Y. Acad. Sci.* **2009**, *1170*, 615–622. [CrossRef]
92. Ulusoy, A.; Rusconi, R.; Perez-Revuelta, B.I.; Musgrove, R.E.; Helwig, M.; Winzen-Reichert, B.; Di Monte, D.A. Caudo-rostral brain spreading of alpha-synuclein through vagal connections. *EMBO Mol. Med.* **2013**, *5*, 1119–1127. [CrossRef] [PubMed]
93. Holmqvist, S.; Chutna, O.; Bousset, L.; Aldrin-Kirk, P.; Li, W.; Bjorklund, T.; Wang, Z.Y.; Roybon, L.; Melki, R.; Li, J.Y. Direct evidence of Parkinson pathology spread from the gastrointestinal tract to the brain in rats. *Acta Neuropathol.* **2014**, *128*, 805–820. [CrossRef]
94. Sampson, T.R.; Debelius, J.W.; Thron, T.; Janssen, S.; Shastri, G.G.; Ilhan, Z.E.; Challis, C.; Schretter, C.E.; Rocha, S.; Gradinaru, V.; et al. Gut Microbiota Regulate Motor Deficits and Neuroinflammation in a Model of Parkinson's Disease. *Cell* **2016**, *167*, 1469–1480. [CrossRef] [PubMed]
95. Shannon, K.; Vanden Berghe, P. The enteric nervous system in PD: Gateway, bystander victim, or source of solutions. *Cell Tissue Res.* **2018**, *373*, 313–326. [CrossRef] [PubMed]
96. Braak, H.; de Vos, R.A.; Bohl, J.; Del Tredici, K. Gastric alpha-synuclein immunoreactive inclusions in Meissner's and Auerbach's plexuses in cases staged for Parkinson's disease-related brain pathology. *Neurosci. Lett.* **2006**, *396*, 67–72. [CrossRef] [PubMed]
97. Van Den Berge, N.; Ferreira, N.; Gram, H.; Mikkelsen, T.W.; Alstrup, A.K.O.; Casadei, N.; Tsung-Pin, P.; Riess, O.; Nyengaard, J.R.; Tamguney, G.; et al. Evidence for bidirectional and trans-synaptic parasympathetic and sympathetic propagation of alpha-synuclein in rats. *Acta Neuropathol.* **2019**, *138*, 535–550. [CrossRef] [PubMed]
98. Kordower, J.H.; Chu, Y.; Hauser, R.A.; Freeman, T.B.; Olanow, C.W. Lewy body-like pathology in long-term embryonic nigral transplants in Parkinson's disease. *Nat. Med.* **2008**, *14*, 504–506. [CrossRef] [PubMed]
99. Li, J.Y.; Englund, E.; Holton, J.L.; Soulet, D.; Hagell, P.; Lees, A.J.; Lashley, T.; Quinn, N.P.; Rehncrona, S.; Bjorklund, A.; et al. Lewy bodies in grafted neurons in subjects with Parkinson's disease suggest host-to-graft disease propagation. *Nat. Med.* **2008**, *14*, 501–503. [CrossRef] [PubMed]
100. Beach, T.G.; White, C.L., 3rd; Hladik, C.L.; Sabbagh, M.N.; Connor, D.J.; Shill, H.A.; Sue, L.I.; Sasse, J.; Bachalakuri, J.; Henry-Watson, J.; et al. Olfactory bulb alpha-synucleinopathy has high specificity and sensitivity for Lewy body disorders. *Acta Neuropathol.* **2009**, *117*, 169–174. [CrossRef] [PubMed]
101. McKinnon, J.; Evidente, V.; Driver-Dunckley, E.; Premkumar, A.; Hentz, J.; Shill, H.; Sabbagh, M.; Caviness, J.; Connor, D.; Adler, C. Olfaction in the elderly: A cross-sectional analysis comparing Parkinson's disease with controls and other disorders. *Int. J. Neurosci.* **2010**, *120*, 36–39. [CrossRef] [PubMed]
102. Haugen, J.; Muller, M.L.; Kotagal, V.; Albin, R.L.; Koeppe, R.A.; Scott, P.J.; Frey, K.A.; Bohnen, N.I. Prevalence of impaired odor identification in Parkinson disease with imaging evidence of nigrostriatal denervation. *J. Neural Transm. (Vienna)* **2016**, *123*, 421–424. [CrossRef]
103. Doty, R.L. Olfactory dysfunction in Parkinson disease. *Nat Rev Neurol* **2012**, *8*, 329–339. [CrossRef] [PubMed]
104. Doty, R.L. Olfaction in Parkinson's disease and related disorders. *Neurobiol. Dis.* **2012**, *46*, 527–552. [CrossRef] [PubMed]
105. Wakabayashi, K.; Takahashi, H.; Takeda, S.; Ohama, E.; Ikuta, F. Parkinson's disease: The presence of Lewy bodies in Auerbach's and Meissner's plexuses. *Acta Neuropathol.* **1988**, *76*, 217–221. [CrossRef] [PubMed]

106. Shannon, K.M.; Keshavarzian, A.; Mutlu, E.; Dodiya, H.B.; Daian, D.; Jaglin, J.A.; Kordower, J.H. Alpha-synuclein in colonic submucosa in early untreated Parkinson's disease. *Mov. Disord.* **2012**, *27*, 709–715. [CrossRef]
107. Killinger, B.A.; Madaj, Z.; Sikora, J.W.; Rey, N.; Haas, A.J.; Vepa, Y.; Lindqvist, D.; Chen, H.; Thomas, P.M.; Brundin, P.; et al. The vermiform appendix impacts the risk of developing Parkinson's disease. *Sci. Transl. Med.* **2018**, *10*. [CrossRef] [PubMed]
108. Savica, R.; Carlin, J.M.; Grossardt, B.R.; Bower, J.H.; Ahlskog, J.E.; Maraganore, D.M.; Bharucha, A.E.; Rocca, W.A. Medical records documentation of constipation preceding Parkinson disease: A case-control study. *Neurology* **2009**, *73*, 1752–1758. [CrossRef]
109. Mulak, A.; Bonaz, B. Brain-gut-microbiota axis in Parkinson's disease. *World J. Gastroenterol.* **2015**, *21*, 10609–10620. [CrossRef] [PubMed]
110. Svensson, E.; Horvath-Puho, E.; Thomsen, R.W.; Djurhuus, J.C.; Pedersen, L.; Borghammer, P.; Sorensen, H.T. Vagotomy and subsequent risk of Parkinson's disease. *Ann. Neurol.* **2015**, *78*, 522–529. [CrossRef] [PubMed]
111. Liu, B.; Fang, F.; Pedersen, N.L.; Tillander, A.; Ludvigsson, J.F.; Ekbom, A.; Svenningsson, P.; Chen, H.; Wirdefeldt, K. Vagotomy and Parkinson disease: A Swedish register-based matched-cohort study. *Neurology* **2017**, *88*, 1996–2002. [CrossRef]
112. Lu, H.T.; Shen, Q.Y.; Xie, D.; Zhao, Q.Z.; Xu, Y.M. Lack of association between appendectomy and Parkinson's disease: A systematic review and meta-analysis. *Aging Clin. Exp. Res.* **2020**, *32*, 2201–2209. [CrossRef] [PubMed]
113. Rey, N.L.; George, S.; Steiner, J.A.; Madaj, Z.; Luk, K.C.; Trojanowski, J.Q.; Lee, V.M.; Brundin, P. Spread of aggregates after olfactory bulb injection of alpha-synuclein fibrils is associated with early neuronal loss and is reduced long term. *Acta Neuropathol.* **2018**, *135*, 65–83. [CrossRef] [PubMed]
114. Rey, N.L.; Steiner, J.A.; Maroof, N.; Luk, K.C.; Madaj, Z.; Trojanowski, J.Q.; Lee, V.M.; Brundin, P. Widespread transneuronal propagation of alpha-synucleinopathy triggered in olfactory bulb mimics prodromal Parkinson's disease. *J. Exp. Med.* **2016**, *213*, 1759–1778. [CrossRef] [PubMed]
115. Uemura, N.; Uemura, M.T.; Lo, A.; Bassil, F.; Zhang, B.; Luk, K.C.; Lee, V.M.; Takahashi, R.; Trojanowski, J.Q. Slow Progressive Accumulation of Oligodendroglial Alpha-Synuclein (alpha-Syn) Pathology in Synthetic alpha-Syn Fibril-Induced Mouse Models of Synucleinopathy. *J. Neuropathol. Exp. Neurol.* **2019**, *78*, 877–890. [CrossRef]
116. Manfredsson, F.P.; Luk, K.C.; Benskey, M.J.; Gezer, A.; Garcia, J.; Kuhn, N.C.; Sandoval, I.M.; Patterson, J.R.; O'Mara, A.; Yonkers, R.; et al. Induction of alpha-synuclein pathology in the enteric nervous system of the rat and non-human primate results in gastrointestinal dysmotility and transient CNS pathology. *Neurobiol. Dis.* **2018**, *112*, 106–118. [CrossRef] [PubMed]
117. Uemura, N.; Yagi, H.; Uemura, M.T.; Hatanaka, Y.; Yamakado, H.; Takahashi, R. Inoculation of alpha-synuclein preformed fibrils into the mouse gastrointestinal tract induces Lewy body-like aggregates in the brainstem via the vagus nerve. *Mol. Neurodegener.* **2018**, *13*, 21. [CrossRef] [PubMed]
118. Kim, S.; Kwon, S.H.; Kam, T.I.; Panicker, N.; Karuppagounder, S.S.; Lee, S.; Lee, J.H.; Kim, W.R.; Kook, M.; Foss, C.A.; et al. Transneuronal Propagation of Pathologic alpha-Synuclein from the Gut to the Brain Models Parkinson's Disease. *Neuron* **2019**, *103*, 627–641. [CrossRef]
119. Uemura, N.; Yagi, H.; Uemura, M.T.; Yamakado, H.; Takahashi, R. Limited spread of pathology within the brainstem of alpha-synuclein BAC transgenic mice inoculated with preformed fibrils into the gastrointestinal tract. *Neurosci. Lett.* **2020**, *716*, 134651. [CrossRef] [PubMed]
120. Burke, R.E.; Dauer, W.T.; Vonsattel, J.P. A critical evaluation of the Braak staging scheme for Parkinson's disease. *Ann. Neurol.* **2008**, *64*, 485–491. [CrossRef]
121. Parkkinen, L.; Hartikainen, P.; Alafuzoff, I. Abundant glial alpha-synuclein pathology in a case without overt clinical symptoms. *Clin. Neuropathol.* **2007**, *26*, 276–283. [CrossRef] [PubMed]
122. Jellinger, K.A. A critical evaluation of current staging of alpha-synuclein pathology in Lewy body disorders. *Biochim. Biophys. Acta* **2009**, *1792*, 730–740. [CrossRef] [PubMed]
123. Gaig, C.; Marti, M.J.; Ezquerra, M.; Cardozo, A.; Rey, M.J.; Tolosa, E. G2019S LRRK2 mutation causing Parkinson's disease without Lewy bodies. *BMJ Case Rep.* **2009**, *2009*. [CrossRef]
124. Chittoor-Vinod, V.G.; Nichols, R.J.; Schule, B. Genetic and Environmental Factors Influence the Pleomorphy of LRRK2 Parkinsonism. *Int. J. Mol. Sci.* **2021**, *22*, 1045. [CrossRef] [PubMed]
125. Johansen, K.K.; Torp, S.H.; Farrer, M.J.; Gustavsson, E.K.; Aasly, J.O. A Case of Parkinson's Disease with No Lewy Body Pathology due to a Homozygous Exon Deletion in Parkin. *Case Rep. Neurol. Med.* **2018**, *2018*, 6838965. [CrossRef] [PubMed]
126. Buchman, A.S.; Shulman, J.M.; Nag, S.; Leurgans, S.E.; Arnold, S.E.; Morris, M.C.; Schneider, J.A.; Bennett, D.A. Nigral pathology and parkinsonian signs in elders without Parkinson disease. *Ann. Neurol.* **2012**, *71*, 258–266. [CrossRef] [PubMed]
127. Braak, H.; Muller, C.M.; Rub, U.; Ackermann, H.; Bratzke, H.; de Vos, R.A.; Del Tredici, K. Pathology associated with sporadic Parkinson's disease–where does it end? *J. Neural Transm. Suppl.* **2006**, 89–97. [CrossRef]
128. Milber, J.M.; Noorigian, J.V.; Morley, J.F.; Petrovitch, H.; White, L.; Ross, G.W.; Duda, J.E. Lewy pathology is not the first sign of degeneration in vulnerable neurons in Parkinson disease. *Neurology* **2012**, *79*, 2307–2314. [CrossRef] [PubMed]
129. Beach, T.G.; Adler, C.H.; Sue, L.I.; Vedders, L.; Lue, L.; White Iii, C.L.; Akiyama, H.; Caviness, J.N.; Shill, H.A.; Sabbagh, M.N.; et al. Multi-organ distribution of phosphorylated alpha-synuclein histopathology in subjects with Lewy body disorders. *Acta Neuropathol.* **2010**, *119*, 689–702. [CrossRef]
130. Adler, C.H.; Beach, T.G. Neuropathological basis of nonmotor manifestations of Parkinson's disease. *Mov. Disord.* **2016**, *31*, 1114–1119. [CrossRef]

131. Sorrentino, Z.A.; Brooks, M.M.T.; Hudson, V., 3rd; Rutherford, N.J.; Golde, T.E.; Giasson, B.I.; Chakrabarty, P. Intrastriatal injection of alpha-synuclein can lead to widespread synucleinopathy independent of neuroanatomic connectivity. *Mol. Neurodegener.* **2017**, *12*, 40. [CrossRef]
132. Bernis, M.E.; Babila, J.T.; Breid, S.; Wusten, K.A.; Wullner, U.; Tamguney, G. Prion-like propagation of human brain derived alpha-synuclein in transgenic mice expressing human wild-type alpha-synuclein. *Acta Neuropathol. Commun.* **2015**, *3*, 75. [CrossRef] [PubMed]
133. Mendez, I.; Vinuela, A.; Astradsson, A.; Mukhida, K.; Hallett, P.; Robertson, H.; Tierney, T.; Holness, R.; Dagher, A.; Trojanowski, J.Q.; et al. Dopamine neurons implanted into people with Parkinson's disease survive without pathology for 14 years. *Nat. Med.* **2008**, *14*, 507–509. [CrossRef] [PubMed]
134. Brundin, P.; Li, J.Y.; Holton, J.L.; Lindvall, O.; Revesz, T. Research in motion: The enigma of Parkinson's disease pathology spread. *Nat. Rev. Neurosci.* **2008**, *9*, 741–745. [CrossRef] [PubMed]
135. Olanow, C.W.; Savolainen, M.; Chu, Y.; Halliday, G.M.; Kordower, J.H. Temporal evolution of microglia and alpha-synuclein accumulation following foetal grafting in Parkinson's disease. *Brain* **2019**, *142*, 1690–1700. [CrossRef] [PubMed]
136. Walsh, D.M.; Selkoe, D.J. A critical appraisal of the pathogenic protein spread hypothesis of neurodegeneration. *Nat. Rev. Neurosci.* **2016**, *17*, 251–260. [CrossRef] [PubMed]
137. Surmeier, D.J.; Obeso, J.A.; Halliday, G.M. Parkinson's Disease Is Not Simply a Prion Disorder. *J. Neurosci.* **2017**, *37*, 9799–9807. [CrossRef] [PubMed]
138. Stokholm, M.G.; Danielsen, E.H.; Hamilton-Dutoit, S.J.; Borghammer, P. Pathological alpha-synuclein in gastrointestinal tissues from prodromal Parkinson disease patients. *Ann. Neurol.* **2016**, *79*, 940–949. [CrossRef]
139. Peelaerts, W.; Bousset, L.; Van der Perren, A.; Moskalyuk, A.; Pulizzi, R.; Giugliano, M.; Van den Haute, C.; Melki, R.; Baekelandt, V. Alpha-Synuclein strains cause distinct synucleinopathies after local and systemic administration. *Nature* **2015**, *522*, 340–344. [CrossRef]
140. Guo, J.L.; Covell, D.J.; Daniels, J.P.; Iba, M.; Stieber, A.; Zhang, B.; Riddle, D.M.; Kwong, L.K.; Xu, Y.; Trojanowski, J.Q.; et al. Distinct alpha-synuclein strains differentially promote tau inclusions in neurons. *Cell* **2013**, *154*, 103–117. [CrossRef]
141. Vasili, E.; Dominguez-Meijide, A.; Outeiro, T.F. Spreading of alpha-Synuclein and Tau: A Systematic Comparison of the Mechanisms Involved. *Front. Mol. Neurosci.* **2019**, *12*, 107. [CrossRef]
142. Angot, E.; Steiner, J.A.; Lema Tome, C.M.; Ekstrom, P.; Mattsson, B.; Bjorklund, A.; Brundin, P. Alpha-synuclein cell-to-cell transfer and seeding in grafted dopaminergic neurons in vivo. *PLoS ONE* **2012**, *7*, e39465. [CrossRef]
143. Reyes, J.F.; Rey, N.L.; Bousset, L.; Melki, R.; Brundin, P.; Angot, E. Alpha-synuclein transfers from neurons to oligodendrocytes. *Glia* **2014**, *62*, 387–398. [CrossRef] [PubMed]
144. Uversky, V.N. Intrinsically disordered proteins from A to Z. *Int. J. Biochem. Cell Biol.* **2011**, *43*, 1090–1103. [CrossRef] [PubMed]
145. Breydo, L.; Wu, J.W.; Uversky, V.N. Alpha-synuclein misfolding and Parkinson's disease. *Biochim. Biophys. Acta* **2012**, *1822*, 261–285. [CrossRef] [PubMed]
146. Helwig, M.; Klinkenberg, M.; Rusconi, R.; Musgrove, R.E.; Majbour, N.K.; El-Agnaf, O.M.; Ulusoy, A.; Di Monte, D.A. Brain propagation of transduced alpha-synuclein involves non-fibrillar protein species and is enhanced in alpha-synuclein null mice. *Brain* **2016**, *139*, 856–870. [CrossRef] [PubMed]
147. Winner, B.; Jappelli, R.; Maji, S.K.; Desplats, P.A.; Boyer, L.; Aigner, S.; Hetzer, C.; Loher, T.; Vilar, M.; Campioni, S.; et al. In vivo demonstration that alpha-synuclein oligomers are toxic. *Proc. Natl. Acad. Sci. USA* **2011**, *108*, 4194–4199. [CrossRef] [PubMed]
148. Chen, S.W.; Drakulic, S.; Deas, E.; Ouberai, M.; Aprile, F.A.; Arranz, R.; Ness, S.; Roodveldt, C.; Guilliams, T.; De-Genst, E.J.; et al. Structural characterization of toxic oligomers that are kinetically trapped during alpha-synuclein fibril formation. *Proc. Natl. Acad. Sci. USA* **2015**, *112*, E1994–E2003. [CrossRef]
149. Cremades, N.; Cohen, S.I.; Deas, E.; Abramov, A.Y.; Chen, A.Y.; Orte, A.; Sandal, M.; Clarke, R.W.; Dunne, P.; Aprile, F.A.; et al. Direct observation of the interconversion of normal and toxic forms of alpha-synuclein. *Cell* **2012**, *149*, 1048–1059. s
150. Meng, F.; Abedini, A.; Plesner, A.; Verchere, C.B.; Raleigh, D.P. The flavanol (-)-epigallocatechin 3-gallate inhibits amyloid formation by islet amyloid polypeptide, disaggregates amyloid fibrils, and protects cultured cells against IAPP-induced toxicity. *Biochemistry* **2010**, *49*, 8127–8133. [CrossRef]
151. Froula, J.M.; Castellana-Cruz, M.; Anabtawi, N.M.; Camino, J.D.; Chen, S.W.; Thrasher, D.R.; Freire, J.; Yazdi, A.A.; Fleming, S.; Dobson, C.M.; et al. Defining alpha-synuclein species responsible for Parkinson's disease phenotypes in mice. *J. Biol. Chem.* **2019**, *294*, 10392–10406. [CrossRef]
152. Gao, X.; Carroni, M.; Nussbaum-Krammer, C.; Mogk, A.; Nillegoda, N.B.; Szlachcic, A.; Guilbride, D.L.; Saibil, H.R.; Mayer, M.P.; Bukau, B. Human Hsp70 Disaggregase Reverses Parkinson's-Linked alpha-Synuclein Amyloid Fibrils. *Mol. Cell* **2015**, *59*, 781–793. [CrossRef] [PubMed]
153. Taguchi, Y.V.; Gorenberg, E.L.; Nagy, M.; Thrasher, D.; Fenton, W.A.; Volpicelli-Daley, L.; Horwich, A.L.; Chandra, S.S. Hsp110 mitigates alpha-synuclein pathology in vivo. *Proc. Natl. Acad. Sci. USA* **2019**, *116*, 24310–24316. [CrossRef] [PubMed]
154. Buell, A.K.; Galvagnion, C.; Gaspar, R.; Sparr, E.; Vendruscolo, M.; Knowles, T.P.; Linse, S.; Dobson, C.M. Solution conditions determine the relative importance of nucleation and growth processes in alpha-synuclein aggregation. *Proc. Natl. Acad. Sci. USA* **2014**, *111*, 7671–7676. [CrossRef] [PubMed]
155. Chakroun, T.; Evsyukov, V.; Nykanen, N.P.; Hollerhage, M.; Schmidt, A.; Kamp, F.; Ruf, V.C.; Wurst, W.; Rosler, T.W.; Hoglinger, G.U. Alpha-synuclein fragments trigger distinct aggregation pathways. *Cell Death Dis.* **2020**, *11*, 84. [CrossRef] [PubMed]

156. Guan, Y.; Zhao, X.; Liu, F.; Yan, S.; Wang, Y.; Du, C.; Cui, X.; Li, R.; Zhang, C.X. Pathogenic Mutations Differentially Regulate Cell-to-Cell Transmission of alpha-Synuclein. *Front. Cell. Neurosci.* **2020**, *14*, 159. [CrossRef] [PubMed]
157. Gegg, M.E.; Verona, G.; Schapira, A.H.V. Glucocerebrosidase deficiency promotes release of alpha-synuclein fibrils from cultured neurons. *Hum. Mol. Genet.* **2020**, *29*, 1716–1728. [CrossRef] [PubMed]
158. Irwin, D.J.; Abrams, J.Y.; Schonberger, L.B.; Leschek, E.W.; Mills, J.L.; Lee, V.M.; Trojanowski, J.Q. Evaluation of potential infectivity of Alzheimer and Parkinson disease proteins in recipients of cadaver-derived human growth hormone. *JAMA Neurol.* **2013**, *70*, 462–468. [CrossRef]
159. Brundin, P.; Melki, R. Prying into the Prion Hypothesis for Parkinson's Disease. *J. Neurosci.* **2017**, *37*, 9808–9818. [CrossRef]
160. Mittal, S.; Bjornevik, K.; Im, D.S.; Flierl, A.; Dong, X.; Locascio, J.J.; Abo, K.M.; Long, E.; Jin, M.; Xu, B.; et al. Beta2-Adrenoreceptor is a regulator of the alpha-synuclein gene driving risk of Parkinson's disease. *Science* **2017**, *357*, 891–898. [CrossRef] [PubMed]
161. Polinski, N.K.; Volpicelli-Daley, L.A.; Sortwell, C.E.; Luk, K.C.; Cremades, N.; Gottler, L.M.; Froula, J.; Duffy, M.F.; Lee, V.M.Y.; Martinez, T.N.; et al. Best Practices for Generating and Using Alpha-Synuclein Pre-Formed Fibrils to Model Parkinson's Disease in Rodents. *J. Parkinsons Dis.* **2018**, *8*, 303–322. [CrossRef] [PubMed]
162. Luk, K.C.; Covell, D.J.; Kehm, V.M.; Zhang, B.; Song, I.Y.; Byrne, M.D.; Pitkin, R.M.; Decker, S.C.; Trojanowski, J.Q.; Lee, V.M. Molecular and Biological Compatibility with Host Alpha-Synuclein Influences Fibril Pathogenicity. *Cell Rep.* **2016**, *16*, 3373–3387. [CrossRef]
163. Devic, I.; Hwang, H.; Edgar, J.S.; Izutsu, K.; Presland, R.; Pan, C.; Goodlett, D.R.; Wang, Y.; Armaly, J.; Tumas, V.; et al. Salivary alpha-synuclein and DJ-1: Potential biomarkers for Parkinson's disease. *Brain* **2011**, *134*, e178. [CrossRef] [PubMed]
164. El-Agnaf, O.M.; Salem, S.A.; Paleologou, K.E.; Cooper, L.J.; Fullwood, N.J.; Gibson, M.J.; Curran, M.D.; Court, J.A.; Mann, D.M.; Ikeda, S.; et al. Alpha-synuclein implicated in Parkinson's disease is present in extracellular biological fluids, including human plasma. *FASEB J.* **2003**, *17*, 1945–1947. [CrossRef] [PubMed]
165. El-Agnaf, O.M.; Salem, S.A.; Paleologou, K.E.; Curran, M.D.; Gibson, M.J.; Court, J.A.; Schlossmacher, M.G.; Allsop, D. Detection of oligomeric forms of alpha-synuclein protein in human plasma as a potential biomarker for Parkinson's disease. *FASEB J.* **2006**, *20*, 419–425. [CrossRef] [PubMed]
166. Mollenhauer, B.; Cullen, V.; Kahn, I.; Krastins, B.; Outeiro, T.F.; Pepivani, I.; Ng, J.; Schulz-Schaeffer, W.; Kretzschmar, H.A.; McLean, P.J.; et al. Direct quantification of CSF alpha-synuclein by ELISA and first cross-sectional study in patients with neurodegeneration. *Exp. Neurol.* **2008**, *213*, 315–325. [CrossRef] [PubMed]
167. Wang, X.; Yu, S.; Li, F.; Feng, T. Detection of alpha-synuclein oligomers in red blood cells as a potential biomarker of Parkinson's disease. *Neurosci. Lett.* **2015**, *599*, 115–119. [CrossRef] [PubMed]
168. Llorens, F.; Schmitz, M.; Varges, D.; Kruse, N.; Gotzmann, N.; Gmitterova, K.; Mollenhauer, B.; Zerr, I. Cerebrospinal alpha-synuclein in alpha-synuclein aggregation disorders: Tau/alpha-synuclein ratio as potential biomarker for dementia with Lewy bodies. *J. Neurol.* **2016**, *263*, 2271–2277. [CrossRef]
169. Shahnawaz, M.; Tokuda, T.; Waragai, M.; Mendez, N.; Ishii, R.; Trenkwalder, C.; Mollenhauer, B.; Soto, C. Development of a Biochemical Diagnosis of Parkinson Disease by Detection of alpha-Synuclein Misfolded Aggregates in Cerebrospinal Fluid. *JAMA Neurol.* **2017**, *74*, 163–172. [CrossRef]
170. Simonsen, A.H.; Kuiperij, B.; El-Agnaf, O.M.; Engelborghs, S.; Herukka, S.K.; Parnetti, L.; Rektorova, I.; Vanmechelen, E.; Kapaki, E.; Verbeek, M.; et al. The utility of alpha-synuclein as biofluid marker in neurodegenerative diseases: A systematic review of the literature. *Biomark. Med.* **2016**, *10*, 19–34. [CrossRef]
171. Ahn, K.J.; Paik, S.R.; Chung, K.C.; Kim, J. Amino acid sequence motifs and mechanistic features of the membrane translocation of alpha-synuclein. *J. Neurochem.* **2006**, *97*, 265–279. [CrossRef] [PubMed]
172. Lee, H.J.; Suk, J.E.; Bae, E.J.; Lee, J.H.; Paik, S.R.; Lee, S.J. Assembly-dependent endocytosis and clearance of extracellular alpha-synuclein. *Int. J. Biochem. Cell Biol.* **2008**, *40*, 1835–1849. [CrossRef]
173. Lee, H.J.; Patel, S.; Lee, S.J. Intravesicular localization and exocytosis of alpha-synuclein and its aggregates. *J. Neurosci.* **2005**, *25*, 6016–6024. [CrossRef] [PubMed]
174. Diaz, E.F.; Labra, V.C.; Alvear, T.F.; Mellado, L.A.; Inostroza, C.A.; Oyarzun, J.E.; Salgado, N.; Quintanilla, R.A.; Orellana, J.A. Connexin 43 hemichannels and pannexin-1 channels contribute to the alpha-synuclein-induced dysfunction and death of astrocytes. *Glia* **2019**, *67*, 1598–1619. [CrossRef] [PubMed]
175. Reyes, J.F.; Sackmann, C.; Hoffmann, A.; Svenningsson, P.; Winkler, J.; Ingelsson, M.; Hallbeck, M. Binding of alpha-synuclein oligomers to Cx32 facilitates protein uptake and transfer in neurons and oligodendrocytes. *Acta Neuropathol.* **2019**, *138*, 23–47. [CrossRef] [PubMed]
176. Ulusoy, A.; Musgrove, R.E.; Rusconi, R.; Klinkenberg, M.; Helwig, M.; Schneider, A.; Di Monte, D.A. Neuron-to-neuron alpha-synuclein propagation in vivo is independent of neuronal injury. *Acta Neuropathol. Commun.* **2015**, *3*, 13. [CrossRef] [PubMed]
177. Wang, B.; Underwood, R.; Kamath, A.; Britain, C.; McFerrin, M.B.; McLean, P.J.; Volpicelli-Daley, L.A.; Whitaker, R.H.; Placzek, W.J.; Becker, K.; et al. 14-3-3 Proteins Reduce Cell-to-Cell Transfer and Propagation of Pathogenic alpha-Synuclein. *J. Neurosci.* **2018**, *38*, 8211–8232. [CrossRef] [PubMed]
178. Abounit, S.; Bousset, L.; Loria, F.; Zhu, S.; de Chaumont, F.; Pieri, L.; Olivo-Marin, J.C.; Melki, R.; Zurzolo, C. Tunneling nanotubes spread fibrillar alpha-synuclein by intercellular trafficking of lysosomes. *EMBO J.* **2016**, *35*, 2120–2138. [CrossRef] [PubMed]
179. Loria, F.; Vargas, J.Y.; Bousset, L.; Syan, S.; Salles, A.; Melki, R.; Zurzolo, C. Alpha-Synuclein transfer between neurons and astrocytes indicates that astrocytes play a role in degradation rather than in spreading. *Acta Neuropathol.* **2017**, *134*, 789–808. [CrossRef] [PubMed]

180. Rostami, J.; Holmqvist, S.; Lindstrom, V.; Sigvardson, J.; Westermark, G.T.; Ingelsson, M.; Bergstrom, J.; Roybon, L.; Erlandsson, A. Human Astrocytes Transfer Aggregated Alpha-Synuclein via Tunneling Nanotubes. *J. Neurosci.* **2017**, *37*, 11835–11853. [CrossRef]
181. Bae, E.J.; Ho, D.H.; Park, E.; Jung, J.W.; Cho, K.; Hong, J.H.; Lee, H.J.; Kim, K.P.; Lee, S.J. Lipid peroxidation product 4-hydroxy-2-nonenal promotes seeding-capable oligomer formation and cell-to-cell transfer of alpha-synuclein. *Antioxid. Redox Signal.* **2013**, *18*, 770–783. [CrossRef] [PubMed]
182. Jang, A.; Lee, H.J.; Suk, J.E.; Jung, J.W.; Kim, K.P.; Lee, S.J. Non-classical exocytosis of alpha-synuclein is sensitive to folding states and promoted under stress conditions. *J. Neurochem.* **2010**, *113*, 1263–1274. [CrossRef] [PubMed]
183. Lee, H.J.; Baek, S.M.; Ho, D.H.; Suk, J.E.; Cho, E.D.; Lee, S.J. Dopamine promotes formation and secretion of non-fibrillar alpha-synuclein oligomers. *Exp. Mol. Med.* **2011**, *43*, 216–222. [CrossRef] [PubMed]
184. Reyes, J.F.; Olsson, T.T.; Lamberts, J.T.; Devine, M.J.; Kunath, T.; Brundin, P. A cell culture model for monitoring alpha-synuclein cell-to-cell transfer. *Neurobiol. Dis.* **2015**, *77*, 266–275. [CrossRef]
185. Courte, J.; Bousset, L.; Boxberg, Y.V.; Villard, C.; Melki, R.; Peyrin, J.M. The expression level of alpha-synuclein in different neuronal populations is the primary determinant of its prion-like seeding. *Sci. Rep.* **2020**, *10*, 4895. [CrossRef]
186. Lee, J.G.; Takahama, S.; Zhang, G.; Tomarev, S.I.; Ye, Y. Unconventional secretion of misfolded proteins promotes adaptation to proteasome dysfunction in mammalian cells. *Nat. Cell Biol.* **2016**, *18*, 765–776. [CrossRef] [PubMed]
187. Xu, Y.; Cui, L.; Dibello, A.; Wang, L.; Lee, J.; Saidi, L.; Lee, J.G.; Ye, Y. DNAJC5 facilitates USP19-dependent unconventional secretion of misfolded cytosolic proteins. *Cell Discov.* **2018**, *4*, 11. [CrossRef] [PubMed]
188. Colombo, M.; Moita, C.; van Niel, G.; Kowal, J.; Vigneron, J.; Benaroch, P.; Manel, N.; Moita, L.F.; Thery, C.; Raposo, G. Analysis of ESCRT functions in exosome biogenesis, composition and secretion highlights the heterogeneity of extracellular vesicles. *J. Cell Sci.* **2013**, *126*, 5553–5565. [CrossRef] [PubMed]
189. Faure, J.; Lachenal, G.; Court, M.; Hirrlinger, J.; Chatellard-Causse, C.; Blot, B.; Grange, J.; Schoehn, G.; Goldberg, Y.; Boyer, V.; et al. Exosomes are released by cultured cortical neurones. *Mol. Cell. Neurosci.* **2006**, *31*, 642–648. [CrossRef] [PubMed]
190. Stuendl, A.; Kunadt, M.; Kruse, N.; Bartels, C.; Moebius, W.; Danzer, K.M.; Mollenhauer, B.; Schneider, A. Induction of alpha-synuclein aggregate formation by CSF exosomes from patients with Parkinson's disease and dementia with Lewy bodies. *Brain* **2016**, *139*, 481–494. [CrossRef] [PubMed]
191. Emmanouilidou, E.; Melachroinou, K.; Roumeliotis, T.; Garbis, S.D.; Ntzouni, M.; Margaritis, L.H.; Stefanis, L.; Vekrellis, K. Cell-produced alpha-synuclein is secreted in a calcium-dependent manner by exosomes and impacts neuronal survival. *J. Neurosci.* **2010**, *30*, 6838–6851. [CrossRef]
192. Alvarez-Erviti, L.; Seow, Y.; Schapira, A.H.; Gardiner, C.; Sargent, I.L.; Wood, M.J.; Cooper, J.M. Lysosomal dysfunction increases exosome-mediated alpha-synuclein release and transmission. *Neurobiol. Dis.* **2011**, *42*, 360–367. [CrossRef] [PubMed]
193. Fussi, N.; Hollerhage, M.; Chakroun, T.; Nykanen, N.P.; Rosler, T.W.; Koeglsperger, T.; Wurst, W.; Behrends, C.; Hoglinger, G.U. Exosomal secretion of alpha-synuclein as protective mechanism after upstream blockage of macroautophagy. *Cell Death Dis.* **2018**, *9*, 757. [CrossRef] [PubMed]
194. Minakaki, G.; Menges, S.; Kittel, A.; Emmanouilidou, E.; Schaeffner, I.; Barkovits, K.; Bergmann, A.; Rockenstein, E.; Adame, A.; Marxreiter, F.; et al. Autophagy inhibition promotes SNCA/alpha-synuclein release and transfer via extracellular vesicles with a hybrid autophagosome-exosome-like phenotype. *Autophagy* **2018**, *14*, 98–119. [CrossRef] [PubMed]
195. Zhang, S.; Eitan, E.; Wu, T.Y.; Mattson, M.P. Intercellular transfer of pathogenic alpha-synuclein by extracellular vesicles is induced by the lipid peroxidation product 4-hydroxynonenal. *Neurobiol. Aging* **2018**, *61*, 52–65. [CrossRef] [PubMed]
196. Guo, M.; Wang, J.; Zhao, Y.; Feng, Y.; Han, S.; Dong, Q.; Cui, M.; Tieu, K. Microglial exosomes facilitate alpha-synuclein transmission in Parkinson's disease. *Brain* **2020**, *143*, 1476–1497. [CrossRef] [PubMed]
197. Sung, J.Y.; Kim, J.; Paik, S.R.; Park, J.H.; Ahn, Y.S.; Chung, K.C. Induction of neuronal cell death by Rab5A-dependent endocytosis of alpha-synuclein. *J. Biol. Chem.* **2001**, *276*, 27441–27448. [CrossRef]
198. Hansen, C.; Angot, E.; Bergstrom, A.L.; Steiner, J.A.; Pieri, L.; Paul, G.; Outeiro, T.F.; Melki, R.; Kallunki, P.; Fog, K.; et al. Alpha-Synuclein propagates from mouse brain to grafted dopaminergic neurons and seeds aggregation in cultured human cells. *J. Clin. Investig.* **2011**, *121*, 715–725. [CrossRef] [PubMed]
199. Kayed, R.; Sokolov, Y.; Edmonds, B.; McIntire, T.M.; Milton, S.C.; Hall, J.E.; Glabe, C.G. Permeabilization of lipid bilayers is a common conformation-dependent activity of soluble amyloid oligomers in protein misfolding diseases. *J. Biol. Chem.* **2004**, *279*, 46363–46366. [CrossRef]
200. Jao, C.C.; Hegde, B.G.; Chen, J.; Haworth, I.S.; Langen, R. Structure of membrane-bound alpha-synuclein from site-directed spin labeling and computational refinement. *Proc. Natl. Acad. Sci. USA* **2008**, *105*, 19666–19671. [CrossRef] [PubMed]
201. Tsigelny, I.F.; Sharikov, Y.; Wrasidlo, W.; Gonzalez, T.; Desplats, P.A.; Crews, L.; Spencer, B.; Masliah, E. Role of alpha-synuclein penetration into the membrane in the mechanisms of oligomer pore formation. *FEBS J.* **2012**, *279*, 1000–1013. [CrossRef] [PubMed]
202. Shrivastava, A.N.; Redeker, V.; Fritz, N.; Pieri, L.; Almeida, L.G.; Spolidoro, M.; Liebmann, T.; Bousset, L.; Renner, M.; Lena, C.; et al. Alpha-synuclein assemblies sequester neuronal alpha3-Na+/K+-ATPase and impair Na+ gradient. *EMBO J.* **2015**, *34*, 2408–2423. [CrossRef] [PubMed]
203. Mao, X.; Ou, M.T.; Karuppagounder, S.S.; Kam, T.I.; Yin, X.; Xiong, Y.; Ge, P.; Umanah, G.E.; Brahmachari, S.; Shin, J.H.; et al. Pathological alpha-synuclein transmission initiated by binding lymphocyte-activation gene 3. *Science* **2016**, *353*. [CrossRef] [PubMed]

204. Ihse, E.; Yamakado, H.; van Wijk, X.M.; Lawrence, R.; Esko, J.D.; Masliah, E. Cellular internalization of alpha-synuclein aggregates by cell surface heparan sulfate depends on aggregate conformation and cell type. *Sci. Rep.* **2017**, *7*, 9008. [CrossRef] [PubMed]
205. Hudak, A.; Kusz, E.; Domonkos, I.; Josvay, K.; Kodamullil, A.T.; Szilak, L.; Hofmann-Apitius, M.; Letoha, T. Contribution of syndecans to cellular uptake and fibrillation of alpha-synuclein and tau. *Sci. Rep.* **2019**, *9*, 16543. [CrossRef] [PubMed]

Review

The PINK1-Mediated Crosstalk between Neural Cells and the Underlying Link to Parkinson's Disease

Elvira Pequeno Leites and Vanessa Alexandra Morais *

Instituto de Medicina Molecular-João Lobo Antunes, Faculdade de Lisboa, Universidade de Lisboa, 1649-028 Lisbon, Portugal; elviraleites@medicina.ulisboa.pt
* Correspondence: vmorais@medicina.ulisboa.pt

Abstract: Mitochondrial dysfunction has a fundamental role in the development of idiopathic and familiar forms of Parkinson's disease (PD). The nuclear-encoded mitochondrial kinase PINK1, linked to familial PD, is responsible for diverse mechanisms of mitochondrial quality control, ATP production, mitochondrial-mediated apoptosis and neuroinflammation. The main pathological hallmark of PD is the loss of dopaminergic neurons. However, novel discoveries have brought forward the concept that a disruption in overall brain homeostasis may be the underlying cause of this neurodegeneration disease. To sustain this, astrocytes and microglia cells lacking PINK1 have revealed increased neuroinflammation and deficits in physiological roles, such as decreased wound healing capacity and ATP production, which clearly indicate involvement of these cells in the physiopathology of PD. PINK1 executes vital functions within mitochondrial regulation that have a detrimental impact on the development and progression of PD. Hence, in this review, we aim to broaden the horizon of PINK1-mediated phenotypes occurring in neurons, astrocytes and microglia and, ultimately, highlight the importance of the crosstalk between these neural cells that is crucial for brain homeostasis.

Keywords: Parkinson's disease; mitochondrial dysfunction; PINK1; neurons; astrocytes; microglia

Citation: Leites, E.P.; Morais, V.A. The PINK1-Mediated Crosstalk between Neural Cells and the Underlying Link to Parkinson's Disease. *Cells* **2021**, *10*, 1395. https://doi.org/10.3390/cells10061395

Academic Editor: Cristine Alves Da Costa

Received: 3 April 2021
Accepted: 31 May 2021
Published: 5 June 2021

1. Introduction

1.1. Parkinson's Disease

Parkinson's disease (PD) is a progressive neurodegenerative movement disorder, mainly characterized by the loss of dopaminergic neurons and the presence of Lewy bodies [1]. Environmental and genetic factors are high contributors to the appearance of this disorder [2]. Non-genetic risk factors include aging, life habits such as smoking, drug abuse and exposure to pesticides, herbicides, and heavy metals [2]. However, a recent study demonstrated that smoking could be a protective factor of PD [3]. On the other hand, genetic mutations have been linked to PD and explain several of the features associated with this pathology [4]. Some of the most prevalently PD-linked mutations are encountered in the genes encoding for α-synuclein (SNCA), leucine-rich repeat kinase 2 (LRRK2), Parkin, PTEN-induced putative kinase (PINK1) and DJ-1 [1].

1.2. Mitochondria and Their Role in PD

Without a doubt, mitochondrial function is crucial for well-being. Hence, the malfunction of this organelle appears associated with multiple diseases, from neurodegeneration to muscle degeneration, among many more [5]. Additionally to the conventional role of ATP production, mitochondria are responsible for calcium homeostasis, apoptosis and reactive oxygen species (ROS) production [6]. All mitochondria-mediated functions are regulated through well-synchronized pathways. These pathways go from mitochondrial dynamics, such as fusion and fission, transport and arrest, all the way to mitochondrial morphology and cristae remodeling. Furthermore, within these synchronized pathways, the formation of mitochondrial-derived vesicles (MDVs) and mitochondria clearance, also known as mitophagy [6]. Knowing that mitochondria are directly associated to these vital cellular processes, it comes as no surprise the impact that their malfunction may have upon cell survival.

The brain is mainly constituted by neurons, astrocytes, microglia and oligodendrocytes. These neural cells have different functions and requirements. Thus one would suspect that mitochondria also would have defined roles for each of these cell types depending on their necessities [7]. In neurons, it has been shown that mitochondria are important for axonal development and regeneration, as they fulfill the local ATP and calcium requirements [7]. These two functions are also required to support synaptic function and plasticity, where ATP and calcium are needed for synaptic-vesicles pool formation and release, respectively [7]. On the other hand, mitochondria found in astrocytes have been shown to be key in regulating glutamate transporters [8]. Astrocytes mainly rely on ATP produced by aerobic glycolysis rather than by oxidative phosphorylation [8]. This cell type is able to produce lactate from pyruvate under aerobic conditions, which may be transmitted to other cells and be used to initiate different pathways such as mitochondrial respiration [9]. The release of lactate from the astrocytes into the extracellular space and its consequent uptake by neurons to perform oxidative phosphorylation, forming the neuron-astrocyte lactate shuttle, further highlights the importance of astrocytes to neuron function [10]. Astrocytes are also able to convert pyruvate into oxaloacetate, allowing the entrance of pyruvate into mitochondria in the absence of α-ketoglutarate [8,11]. This conversion, performed by pyruvate carboxylase, an enzyme highly enriched in astrocytes, will allow the formation of glutamate and eventually, glutamine, which is fundamental for the glutamine-glutamate cycling between neurons and astrocytes [11,12]. Since the transport of glutamate and its precursor glutamine from the blood to the brain is a rather slow process [13], glutamate, an excitatory transmitter, and its decarboxylation product γ-aminobutyric acid (GABA), an inhibitory transmitter, need to be synthesized by neural cells [14]. However, neurons are not able to produce glutamate as they lack the α-ketoglutarate enzyme [14]. Therefore, astrocytes that have high pyruvate carboxylase activity are able to produce a higher amount of oxaloacetate, leading to the formation of more α-ketoglutarate, thus more glutamate [11,15]. This glutamate, together with the neuronal-released glutamate, is metabolized by an astrocyte-specific enzyme called glutamine synthetase, leading to the formation of glutamine, which is transported to neurons and converted into glutamate or GABA, according to the necessity of the neurons [14]. The uptake of glutamate by astrocytes is made via EAATs (excitatory amino acid transports) together with sodium ions, which are then excreted by the action of the Na^+/K^+ ATPase expending ATP [16]. This reaction, mediated by Na^+/K^+ ATPase, leads to glucose uptake from the circulation through the glucose transporter GLUT1, which will be converted into lactate and shuttled to neurons to be used as an energy substrate [16]. A proteomic study, using an engineered MitoTag mouse, revealed that astrocytes had increased expression levels of peroxisomal proteins and enzymes involved in mitochondrial β-oxidation when compared with Purkinje cells and granule cells [17]. Regarding microglia, not much is known about the specifications of mitochondria in these cells in resting conditions. However, when comparing activated with non-activated microglia, studies have revealed that after the activation of microglia, there is a switch from mitochondrial oxidative phosphorylation to anaerobic glycolysis [8,18]. Also, inhibition of Complex I leads to activation of microglia, while deficits in mitochondrial fission pathways reduce activation, showing the importance of mitochondria dynamics in the activation status of the cell [19,20].

Notably, mitochondria are crucial for mediating cell survival and ultimately tissue or organ homeostasis. Therefore, it is not surprising that mitochondrial dysfunction is implicated in several diseases, namely brain-related disorders. For instance, the development of Parkinson-like symptoms has been associated to close and prolonged exposure to pesticides, herbicides and neurotoxins such as MPTP [2]. Rotenone and paraquat are two commonly used pesticides that are known mitochondrial toxins and that lead to dopaminergic neuron loss [21]. Rotenone, an inhibitor of mitochondrial Complex I, and paraquat, which prevents electron transfer to NADPH, were shown to cause oxidative stress by triggering intracellular ROS formation in the striatum [21,22]. Although no direct connection of rotenone and paraquat with PD patients was proven, a study using two groups of pesticides, which inhibit Complex I and increase oxidative stress, showed that prolonged use of these compounds has a positive correlation with the development of PD [21]. MPP^+, a metabolite of MPTP, enters dopaminergic neurons inhibiting mitochondrial respiration by inhibiting Complex I [23]. These studies show that mitochondrial dysfunction can, in principle, be one of the underlying causes of PD.

The identification of mutations in genes that encode Parkin, PINK1 and even mutations in mitochondrial DNA further strengthens the hypothesis that mitochondria are one of the main causes of PD [5]. When studying early-stage PD patients and pre-symptomatic PD patients, which are represented by incidental Lewy body disease cases, mitochondrial DNA (mtDNA) mutations were observed in *substantia nigra* neurons when compared to control samples [24]. Also, when comparing total mtDNA deletions/rearrangements of patients with PD, multiple system atrophy (MSA), dementia with Lewy bodies (DLB), Alzheimer disease (AD) and age-matched controls, the number and variety of mtDNA rearrangements were significantly increased in PD patients' brains [25]. Loss-of-function mutations in Parkin and PINK1 are related to alterations in mitochondrial function either by impairing Ca^{2+} homeostasis and ATP production, by impairing the clearance of damaged mitochondria in a process called mitophagy, by increasing cell apoptosis in a mitochondrial-dependent manner, among a variety of other pathways that are impaired when PINK1 is mutated [26–29].

PINK1, a nuclear-encoded mitochondrial serine/threonine kinase, is a promiscuous kinase as the kinases' substrates are phosphorylated depending on the overall status of the mitochondria [30]. Under healthy conditions, PINK1 is recruited into the mitochondria, where it is cleaved by different proteases [31]. Primarily, PINK1 is cleaved by the mitochondrial processing peptidase (MPP), followed by presenilin-associated rhomboid-like protease (PARL), m-AAA and ClXP [32]. These cleavages mediate the turnover of PINK1, ending with the retro-translocation of PINK1 to the cytosol, where further processing occurs in a proteasome-dependent manner [33]. During the internalization of PINK1 into the mitochondria, proteins such as NDUFA10, TRAP1 (TNF receptor-associated protein 1), BCL-xL and HtrA2 are phosphorylated [34–37]. NDUFA10 is a Complex I subunit, and its phosphorylation mediates the overall enzymatic function of the ETC, and so it is important for ATP production [38]. Acting as a cell survival mechanism, PINK1 phosphorylates BCL-xL inhibiting in pro-apoptotic cleavage [37]. On the other hand, phosphorylation of mitochondrial chaperone TRAP1 and HtrA2 protects cells from mitochondrial-induced apoptosis [35,36]. However, when PINK1 encounters unhealthy mitochondria, the accumulation of full-length PINK1 occurs at the outer mitochondrial membrane, leading to an increase in the recruitment of cytosolic Parkin, followed by the PINK1-mediated phosphorylation of Parkin, ubiquitin and PINK1 itself and giving rise to mitophagy, a mitochondrial-specific clearance pathway [39–41]. Parkin, also a known PINK1 substrate, is a cytosolic ubiquitin E3 ligase known to cause PD [42]. Additionally, when mitochondria contain damaged cargo, this PINK1-Parkin interaction is responsible for the formation of mitochondrial-derived vesicles that were shown to be a delivery mechanism of oxidized cargo to the late endosome [43,44]. The formation of these vesicles depends on the presence of PINK1 and Parkin, and it is a process that precedes mitophagy, indicating that it is a first attempt to rescue mitochondria before initiating mitophagy [43]. The triggering of MDVs also differs from mitophagy. While mitophagy requires a global mitochondrial depolarization, MDVs can be generated with the increase in ROS [43]. However, the molecular mechanism by which mitochondrial-derived vesicles are formed is still not well known.

Although these studies postulate that mitochondrial dysfunction occurs in several forms of PD, the fact that these dysfunctions mainly afflict dopaminergic neurons needs to be clarified.

2. Neurons in PD

Dopaminergic neurons are particularly sensitive to the changes that happen in a PD-afflicted brain, and a progressive malfunction and eventual loss of these neurons lead to the appearance of motor symptoms [1]. Axon length and the level of myelinization are plausible reasons why some neurons are more predisposed to enter apoptosis than others [45]. Other than the fact that the more afflicted neurons have long and thin axons, they are also unmyelinated or only partially myelinated, as previously shown by Braak [45,46]. This could be explained by their extremely high energy turnover and possible consequent exposure to oxidative stress [45]. However, this increase in susceptibility is still not well known. Studies performed in *drosophila* showed that dopaminergic neurons appear more sensitive to oxidative stress, a phenotype that was reverted with the use of antioxidants [47]. Additionally, PINK1 loss-of-function was shown to be implicated in this progressive loss of dopaminergic neurons in *drosophila* as it leads to increasing levels of oxidative stress [47]. Mutations in the *PINK1* gene were associated with familial and sporadic early-onset PD [48,49]. Although genetic mutations are mostly associated with familial forms of PD, *PINK1* mutations were found in an Italian cohort of sporadic patients [49]. However, the *PINK1* gene is not the only PD-related gene associated with sporadic Parkinson's disease. PARKIN and DJ-1 have also been associated with this form of the disorder [49]. For these reasons, it is important to study the impact of PINK1 in neurons under physiological and pathological conditions in order to fully understand how this mitochondrial kinase impacts neuronal loss (Figure 1).

PINK1 in Neurons

PINK1 is a key regulator of mitochondrial quality control. When PINK1 is mutated and is not able to perform its functions in a healthy mitochondrion, ATP production decreases, ROS production increases, increasing neurotoxicity and dopaminergic neuronal death either by the increase of ROS or by the absence of protective pathways [50]. The impact of increased ROS is supported by the protection of dopaminergic neurons when using antioxidants in PINK1-dependent models [47]. The impact of PINK1 downregulation in *drosophila* was accessed by comparing dopaminergic and serotonergic neurons and only dopaminergic neurons suffered progressive neurodegeneration [47]. However dopaminergic neurons were not the only cells affected, as an ommatidial degeneration was also observed [47]. The reason why dopaminergic are more sensitive than other neuronal types to the lack of PINK1 is still an unanswered question.

When mitochondria are damaged, and in order to inhibit its movement to an energy-dependent region in neurons, Miro (mitochondria Rho) is phosphorylated by PINK1, leading to a mitochondrial arrest [51]. Miro was suggested to be an important adaptor for the crosstalk between dynein and kinesin transport, mediating the anterograde and retrograde transport of mitochondria in neurons [51]. However, when mitochondria are damaged, Miro appears to be phosphorylated at Ser^{156} by PINK1 and ubiquitinated by Parkin, inhibiting its action and consequent mitochondrial movement [51,52]. In this situation, fusion should be decreased and fission increased in order to degrade the minimum amount of mitochondria necessary to eliminate the damage. For this, when mitochondria are depolarized, Mfn2 (mitofusin 2) and DRP1 (dynamin-related protein 1), proteins involved in mitochondrial fusion and fission, respectively, are phosphorylated in a PINK1-dependent manner, highlighting the importance of PINK1 in regulating this process [53–55]. Mfn2 is one of the proteins responsible for the fusion of mitochondria. However, when Mfn2 is phosphorylated by PINK1 and consequently ubiquitinated by Parkin, it is degraded, preventing a fusion event [53]. In the case of DRP1, a key player of mitochondrial fission, when phosphorylated at Ser^{616} in a PINK1-dependent fashion, fission is promoted However, no mechanistic insights are known [56]. In PINK1-linked PD, this control and clearance of damaged mitochondria, as well as this fusion and fission balance, is impaired leading to the release of ROS and damaged mitochondrial DNA, both of which are toxic products that increase neurotoxicity and afflict dopaminergic neurons [50,57].

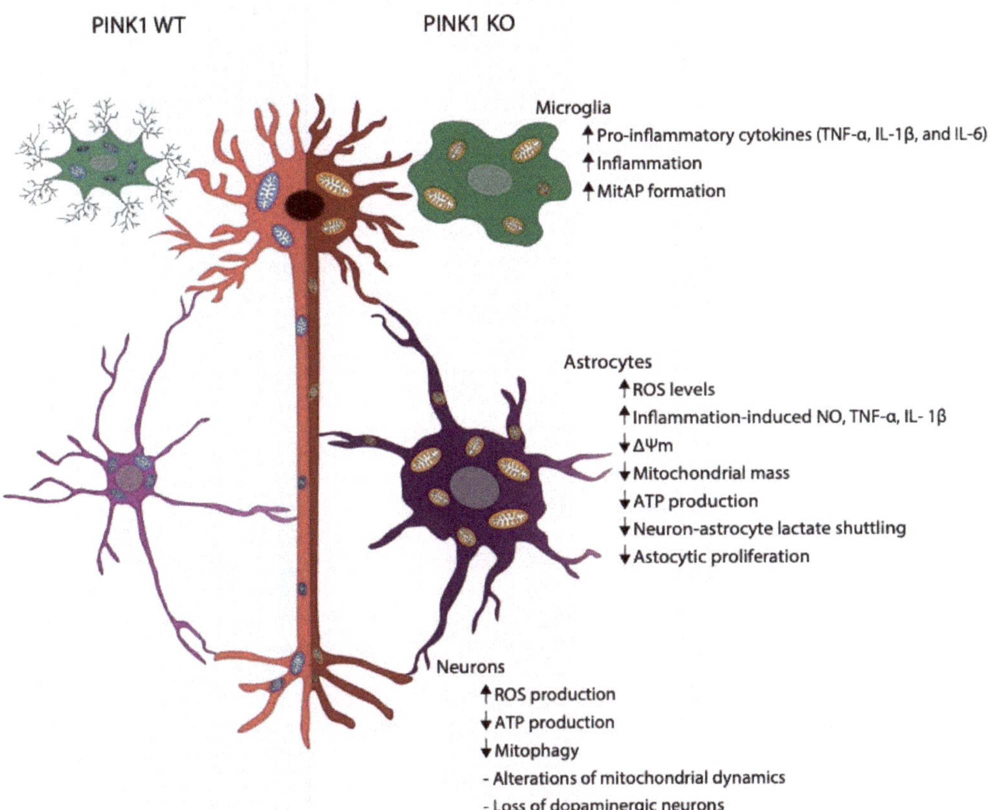

Figure 1. Impact of PINK1 deficiency in neural cells. When compared with PINK1 WT, PINK1 KO microglia shows an increase in the pro-inflammatory release of the cytokines TNF-α, IL-1β and IL-6, consequently leading to an increase in overall brain inflammation. Additionally, in the absence of PINK1, an increase in mitochondrial antigen presentation (MitAP) occurs, indicating an activation of autoimmune mechanisms. In astrocytes lacking PINK1, an increase in reactive oxygen species (ROS), inflammation-induced nitric oxide (NO) levels and TNF-α and IL-1β production has been observed, while a decrease in mitochondrial membrane potential (Δψm), mitochondrial mass, ATP production and glucose-uptake capacity occurs. These alterations prime a decreased astrocytic proliferation ability. In neurons, PINK1 loss-of-function leads to a decrease in ATP production and mitochondrial clearance, as well as an increase in ROS production. Absence of PINK1 also alters mitochondrial dynamics. All these alterations lead to an ultimate loss of dopaminergic neurons.

PINK1 and Parkin were shown to regulate mitochondrial biogenesis and to maintain a pool of healthy mitochondria in dopaminergic neurons through the PARIS/PGC-1α axis [58]. However, when PINK1 or Parkin are defective, a progressive dopaminergic neuron loss occurs, demonstrating another pathway where PINK1's loss-of-function could be the cause of PD [57,58].

As previously mentioned, mitochondria are responsible for calcium homeostasis, and PINK1 also regulates this mechanism, as proven by the impairment of mitochondrial calcium efflux and consequent mitochondrial calcium overload in the absence of PINK1 [59]. This calcium dysregulation results in increased ROS levels in PINK1 KO mouse neurons, leading to an impaired respiration and mitochondrial permeability transition pore (PTP) opening, ultimately promoting neuronal death [59]. This is of particular importance for neurons as they are more susceptible to calcium influxes and increased oxidative stress, as in the case of neurons from the *substantia nigra* [59].

The loss of dopaminergic neurons is a pathological hallmark of PD. However, even though PINK1 is present in all cells of the body, only the neurons from PD patients are afflicted [60]. When looking into PINK1 function in other organs, such as kidney, PINK1-mediated mitophagy has a protective role, preventing renal tubular epithelial cells apoptosis and tissue damage in contrast-induced acute kidney injury by reducing mitochondrial ROS and neutrophil/lymphocyte ratio family pyrin domain containing 3 (NLRP3) inflammasome activation, in mice [61]. In mice livers, PINK1-mediated mitophagy was shown to have a protective role against non-alcoholic fatty liver disease (NAFLD) by clearing damaged mitochondria and allowing cyanidin-3-O-glucoside (C3G) to suppress oxidative stress, NLRP3 inflammasome activation and improving glucose metabolism [62]. In adult mouse cardiomyocytes, phosphorylation of PINK1 at Ser^{495}, by AMP-activated protein kinase $\alpha2$ (AMPK$\alpha2$), was shown to increase mitophagy after stimulation, decreasing ROS production and apoptosis of cardiomyocytes demonstrating a role in preventing the progression of heart failure [63]. Taking these protective roles in different organs and diseases is not surprising that according to different insults and different environments, PINK1 has different functions and significance. The increase in sensitivity of dopaminergic neurons to the absence of PINK1 is still not known. However, one could argue that instead of being more sensitive to the absence of PINK1, these neurons could be more sensitive to changes in their environment that are caused by the lack of PINK1-mediated mitochondrial quality control. While PD patients age, they are exposed to different diseases, such as bacterial or viral infections. These changes in the body's immunity could be an explanation to why PD patients develop symptoms after some years, even when they have PINK1 mutations since birth, as shown in the study where a bacterial infection was enough to induce PD-like symptoms in mice lacking PINK1 [64]. With this stimulus, microglia and astrocytes lacking PINK1 may not be able to restore their physiological function and support neurons. For these reasons, and as neurons are sensitive to a homeostatic environment in order to maintain their function and plasticity, it is important to decipher the impact that PINK1 loss-of-function causes in different cell types.

3. Astrocytes in PD

Neurons need to be in contact with functioning astrocytes in order to maintain synaptic homeostasis, local blood flow and neural network activity [65]. Astrocytes are the most populous sub-type of glial cells in the brain [66]. In addition to the main functions already mentioned above, the importance of astrocyte to dopaminergic neurons survival was further underlined when the protective function of GDNF (glial-derived neurotrophic factor), one of the neurotrophic molecules released by astrocytes, was observed [67]. Neuroinflammation is a well-demonstrated characteristic of PD [68]. This process can be mediated either by the activation of astrocytes or microglia [69,70]. Results obtained using PD-patient iPSC-derived astrocytes showed that α-synuclein also accumulates in these cells leading to an impairment in chaperone-mediated autophagy that increases the accumulation of α-synuclein resulting in non-cell-autonomous neurodegeneration [71]. This ability of astrocytes to uptake α-synuclein, decreasing its toxicity towards neurons, leads to an increase in intracellular toxic deposits of α-synuclein in astrocytes, consequently resulting in mitochondrial damage reflected by the presence of fragmented mitochondria and an overall decrease in ATP content [72]. Astrocytes also have the ability to keep neuronal homeostasis by taking up cellular debris or other toxic material that can be released from neighboring cells [73]. This feature is also important at the beginning of PD as it will reduce inflammation and also during the development of the disease as the death of dopaminergic neurons occurs. Recently, it was shown that astrocytes have the capacity to degrade dysfunctional mitochondria that originated from afflicted dopaminergic neurons, and consequently by providing healthy mitochondria to neurons, revealing once again the importance of neuron-astrocyte communication [74,75].

Furthermore, since astrocytes provide energy to neurons, mitochondrial dysfunction can also have a major impact in neuronal survival [8]. In accordance with this fact, PINK1 loss-of-function has started to be investigated in astrocytes.

PINK1 in Astrocytes

It has been reported that PINK1 expression increases in astrocytes during mouse brain development and that PINK1 levels can affect the number of glial fibrillary acidic protein (GFAP)-positive astrocytes, GFAP being a widely-used protein maker for astrocytes [76]. Choi and co-workers conclude that PINK1 is a crucial protein for the development and function of astrocytes. However, the molecular mechanism remains elusive (Figure 1) [76]. As previously mentioned, in the presence of dysfunctional mitochondria, PINK1 phosphorylates Parkin and ubiquitin [40]. Even though ubiquitin phosphorylation by PINK1 is increased in astrocytes, when compared with other neural types, the physiological explanation for this event is not yet known [77]. Since PINK1 is so important for maintaining a healthy pool of mitochondria, it is not surprising that its absence in astrocytes leads to mitochondrial defects, such as decreased mitochondrial membrane potential, mitochondrial mass, increased ROS levels, decreased ATP production and decreased glucose-uptake ability [78]. All these mitochondrial phenotypes lead to a decreased proliferation of astrocytes, consequently leading to decreased wound healing capacity, as well as all other basal functions of astrocytes [78]. In the absence of PINK1, astrocytes stimulated with lipopolysaccharide (LPS) and interferon-γ (IFN-γ), present an abnormal innate immune response and increased inflammation-induced nitric oxide (NO), tumor necrosis factor-alpha (TNF-α) and interleukin-1 β (IL-1β) production, a possible mechanism through which neurons die [79].

Although astrocytes mediate inflammation and could, in principle, be responsible for a neurotoxic effect, microglia can also regulate and activate astrocytes by releasing soluble cytokines and chemokines [80]. The mitochondrial-mediated activation of astrocytes can be done by increasing the release of TNF-α and IL-1β by microglia, inducing morphological and biochemical alterations in astrocytes [81]. Having this in mind, microglia is another highly relevant cell type in PINK1-dependent PD.

4. Microglia in PD

Defined as the residing macrophages of the central nervous system, microglia are the most abundant immune cells in the brain [82]. The main function of microglia is to protect the brain from injury [83]. Microglia have to be able to regulate the inflammation either through repair, regeneration or cytotoxicity [83]. Depending on the activation state of microglia, these can either release pro-inflammatory cytokines or neurotoxic molecules that can potentiate the inflammation, being cytotoxic, or produce anti-inflammatory molecules, neurotrophic factors or increasing their engulfment capacity that help to restore homeostasis, promoting repair or regeneration [83]. Although microglia is mainly associated with inflammation, it was shown that in multiple sclerosis, it has an important role in promoting tissue recovery, either by producing protective factors for remyelination, phagocyting apoptotic cells and debris promoting regeneration and proliferation of stem cells, or recruiting oligodendrocytes precursors cells stimulating neurogenesis [84]. A variety of factors, such as duration of the insult, type of insult, interaction with other cell types and even the amount of cytokines released by microglia, will determine if the action of microglia is beneficial or harmful for the brain, and this will be the difference between restoring the homeostasis of the brain or supporting the progression of neurodegenerative disease [83]. Recently more importance has been given to microglia, and besides their pro-inflammatory role, these cells are also able to engulf debris and release anti-inflammatory factors, such as transforming growth factor (TGF)-β or IL-10 [82]. In PD, dopaminergic neurons release aggregates of α-synuclein when entering apoptosis that triggers a microglia-mediated pro-inflammatory behavior [85]. Under physiological conditions, microglia are responsible for synaptic pruning and remodeling, engulfing apoptotic cells and cell debris [86,87].

PET (positron emission tomography) studies performed using PD patients, demonstrate that microglia activation is an early and prolonged response of PD [88]. Also neuroinflammation mediated by IL-1β, which is released by microglia and can activate astrocytes, increases dopamine neurons' susceptibility to death [70]. On the other hand inhibition of astrocytic activation by microglia is neuroprotective in PD models [89].

PINK1 in Microglia

Since microglia have pro and anti-inflammatory functions, it is important to know what happens to this cell type when in the presence of mutated PINK1 (Figure 1).

In PD, it was shown that aggregates of α-synuclein result in reactive pro-inflammatory microglia leading to an increase in TNF-α, NO, and IL-1β [90]. However, in the absence of PINK1, an increase in pro-inflammatory released cytokines (TNF-α, IL-1β, and IL-6) in injured mouse brain slices is observed, suggesting that PINK1 has a protective role [91]. A few years ago, it was shown that PINK1 and Parkin have an important role in adaptive immunity through the repression of MitAP (mitochondrial antigen presentation) [92]. This process occurs in immune cells, and is stimulated by inflammatory conditions, suggesting that PINK1 also has an impact on immunity [92]. After this discovery, it was shown that a Gram-negative bacterial infection in the intestines of PINK1 knock-out mice increases MitAP and autoimmune mechanisms leading to a decrease in dopaminergic neuron density [64]. Since microglia are the resident macrophages of the brain and inflammation is a marked feature of PD, the impact that PINK1's loss-of-function in microglia needs to be clarified. However, the previously described functions of PINK1 acting as a mitochondrial quality control regulator should not be discarded when considering the overall well-being of microglia and the crucial role at keeping dopaminergic neurons healthy and in a low inflammatory environment.

5. PINK1, a Putative Mediator of the Crosstalk between Neural Cells

Even though there are other therapeutic approaches under development and continuous clinical trials ongoing, such as gene therapies, immunotherapies targeting α-synuclein or stem cell-based treatments, levodopa is at present the most commonly used treatment for PD patients as it significantly reduces motor symptoms [93,94]. However, and in order to develop novel treatments for PD, it is important to decipher the molecular mechanisms responsible for neuronal death and ultimately disease progression by taking into account overall brain homeostasis. Crosstalk between neurons, astrocytes and microglia is becoming more evident. The sensitivity of dopaminergic neurons to impaired environmental homeostasis appears as one of the main causes of PD, the maintenance of this homeostasis is the responsibility of the astrocytes and the microglia [1]. PINK1 is a key player for maintaining mitochondrial fitness [6]. For this reason, it is crucial to unravel the specific impact that PINK1 mutations have upon these three neural cell types.

In this review, we describe different phenotypes mediated by the absence of PINK1 that, independently of its localization, be it in neurons, astrocytes or microglia, lead to PD. The homeostasis required for an adequate function and survival of neurons is disrupted when PINK1 is not able to perform properly in astrocytes or microglia. An increase in inflammation, mediated by microglia and astrocytes, for a long period of time is not beneficial for neuron survival, ultimately leading to neuronal dysfunction and death [79,88]. The decreased ATP production by astrocytes lacking PINK1 could potentially affect the overall energy levels of neurons, resulting in neuronal deficit and increased ROS production, which will activate microglia and initiate an inflammatory response. Additionally, impaired PINK1 present in astrocytes may reduce the ability of astrocytic-mitochondrial transfer to damaged neurons, leading to an accumulation of damaged mitochondria in these cells. Baring this crosstalk between neuron-microglia-astrocyte in mind, the impact of PINK1 loss-of-function is detrimental to maintain a healthy pool of mitochondria within each neural cell type and, ultimately, to regulate efficient and robust bioenergetics crosstalk between these cells.

6. Conclusions

Although dopaminergic neurons are the most affected cells in PD, it has been recently demonstrated that non-neuronal cells, including astrocytes and microglia, can have a crucial role in both idiopathic and inherited forms of the disease [77,87,95]. Additionally, mitochondrial dysfunction in neurons, astrocytes and microglia may have a devastating impact on the function and survival of these cells, hence on overall brain homeostasis. Thus, understanding the molecular mechanism regulated by PINK1 in the brain will aid in gaining knowledge on how overall mitochondrial homeostasis is underlying several PD pathologies.

Author Contributions: Writing, E.P.L. and V.A.M. All authors have read and agreed to the published version of the manuscript.

Funding: This work was supported by the funding references: Fundação para a Ciência e Tecnologia (FCT, PTDC/BIA-CEL/31230/2017); European Molecular Biology Organization (EMBO-IG#3309), and European Research Council Starting Grant (ERC-StG#679168). EPL is a holder of an FCT PhD fellowship (PD/BD/128289/2017). VAM is an iFCT researcher (IF/01693/2014).

Conflicts of Interest: The authors declare no conflict of interest.

References

1. Kouli, A.; Torsney, K.M.; Kuan, W.-L. Parkinson's disease: Etiology, neuropathology, and pathogenesis. In *Parkinson's Disease: Pathogenesis and Clinical Aspects*; Codon Publications: Brisbane, Australia, 2018; pp. 3–26, ISBN 9780994438164.
2. De Lau, L.M.; Breteler, M.M. Epidemiology of Parkinson's disease. *Lancet Neurol.* **2006**, *5*, 525–535. [CrossRef]
3. Mappin-Kasirer, B.; Pan, H.; Lewington, S.; Kizza, J.; Gray, R.; Clarke, R.; Peto, R. Tobacco smoking and the risk of Parkinson disease. *Neurology* **2020**. [CrossRef]
4. Wertz, M.H.; Heiman, M. Genome-wide genetic screening in the mammalian CNS. In *Research and Perspectives in Neurosciences*; Springer: Berlin/Heidelberg, Germany, 2017; pp. 31–39.
5. Morais, V.A.; de Strooper, B. Mitochondria dysfunction and neurodegenerative disorders: Cause or consequence. *J. Alzheimers Dis.* **2010**, *20*, S255–S263. [CrossRef]
6. Leites, E.P.; Morais, V.A. Mitochondrial quality control pathways: PINK1 acts as a gatekeeper. *Biochem. Biophys. Res. Commun.* **2018**, *500*, 45–50. [CrossRef]
7. Rangaraju, V.; Lewis, T.L.; Hirabayashi, Y.; Bergami, M.; Motori, E.; Cartoni, R.; Kwon, S.K.; Courchet, J. Pleiotropic mitochondria: The influence of mitochondria on neuronal development and disease. *J. Neurosci.* **2019**, *39*, 8200–8208. [CrossRef] [PubMed]
8. McAvoy, K.; Kawamata, H. Glial mitochondrial function and dysfunction in health and neurodegeneration. *Mol. Cell. Neurosci.* **2019**, *101*, 103417. [CrossRef]
9. Barros, L.F.; Brown, A.; Swanson, R.A. Glia in brain energy metabolism: A perspective. *Glia* **2018**, *66*, 1134–1137. [CrossRef] [PubMed]
10. Rose, J.; Brian, C.; Woods, J.; Pappa, A.; Panayiotidis, M.I.; Powers, R.; Franco, R. Mitochondrial dysfunction in glial cells: Implications for neuronal homeostasis and survival. *Toxicology* **2017**. [CrossRef] [PubMed]
11. Shank, R.P.; Bennett, G.S.; Freytag, S.O.; Campbell, G.L.M. Pyruvate carboxylase: An astrocyte-specific enzyme implicated in the replenishment of amino acid neurotransmitter pools. *Brain Res.* **1985**. [CrossRef]
12. Gamberino, W.C.; Berkich, D.A.; Lynch, C.J.; Xu, B.; LaNoue, K.F. Role of pyruvate carboxylase in facilitation of synthesis of glutamate and glutamine in cultured astrocytes. *J. Neurochem.* **1997**. [CrossRef] [PubMed]
13. Smith, Q.R. Transport of glutamate and other amino acids at the blood-brain barrier. *J. Nutr.* **2000**, *130*, 1016S–1022S. [CrossRef] [PubMed]
14. Hertz, L.; Rothman, D.L. Glutamine-glutamate cycle flux is similar in cultured astrocytes and brain and both glutamate production and oxidation are mainly catalyzed by aspartate aminotransferase. *Biology* **2017**, *6*, 17. [CrossRef]
15. Yu, A.C.H.; Drejer, J.; Hertz, L.; Schousboe, A. Pyruvate carboxylase activity in primary cultures of astrocytes and neurons. *J. Neurochem.* **1983**. [CrossRef] [PubMed]
16. Bélanger, M.; Allaman, I.; Magistretti, P.J. Brain energy metabolism: Focus on astrocyte-neuron metabolic cooperation. *Cell Metab.* **2011**, *14*, 724–738. [CrossRef]
17. Fecher, C.; Trovò, L.; Müller, S.A.; Snaidero, N.; Wettmarshausen, J.; Heink, S.; Ortiz, O.; Wagner, I.; Kühn, R.; Hartmann, J.; et al. Cell-type-specific profiling of brain mitochondria reveals functional and molecular diversity. *Nat. Neurosci.* **2019**. [CrossRef]
18. Orihuela, R.; McPherson, C.A.; Harry, G.J. Microglial M1/M2 polarization and metabolic states. *Br. J. Pharmacol.* **2016**, *173*, 649–665. [CrossRef]
19. Gao, F.; Chen, D.; Hu, Q.; Wang, G. Rotenone directly induces BV2 cell activation via the p38 MAPK pathway. *PLoS ONE* **2013**, *8*, e72046. [CrossRef]

20. Park, J.; Choi, H.; Min, J.S.; Park, S.J.; Kim, J.H.; Park, H.J.; Kim, B.; Chae, J., II; Yim, M.; Lee, D.S. Mitochondrial dynamics modulate the expression of pro-inflammatory mediators in microglial cells. *J. Neurochem.* **2013**. [CrossRef] [PubMed]
21. Tanner, C.M.; Kame, F.; Ross, G.W.; Hoppin, J.A.; Goldman, S.M.; Korell, M.; Marras, C.; Bhudhikanok, G.S.; Kasten, M.; Chade, A.R.; et al. Rotenone, paraquat, and Parkinson's disease. *Environ. Health Perspect.* **2011**, *119*, 866–872. [CrossRef] [PubMed]
22. Kuter, K.; Nowak, P.; Gołembiowska, K.; Ossowska, K. Increased reactive oxygen species production in the brain after repeated low-dose pesticide paraquat exposure in rats. A comparison with peripheral tissues. *Neurochem. Res.* **2010**. [CrossRef] [PubMed]
23. Krueger, M.J.; Singer, T.P.; Casida, J.E.; Ramsay, R.R. Evidence that the blockade of mitochondrial respiration by the neurotoxin 1-methyl-4-phenylpyridinium (MPP+) involves binding at the same site as the respiratory inhibitor, rotenone. *Biochem. Biophys. Res. Commun.* **1990**, *169*, 123–128. [CrossRef]
24. Lin, M.T.; Cantuti-Castelvetri, I.; Zheng, K.; Jackson, K.E.; Tan, Y.B.; Arzberger, T.; Lees, A.J.; Betensky, R.A.; Beal, M.F.; Simon, D.K. Somatic mitochondrial DNA mutations in early Parkinson and incidental lewy body disease. *Ann. Neurol.* **2012**. [CrossRef]
25. Gu, G.; Reyes, P.F.; Golden, G.T.; Woltjer, R.L.; Hulette, C.; Montine, T.J.; Zhang, J. Mitochondrial DNA deletions/rearrangements in Parkinson disease and related neurodegenerative disorders. *J. Neuropathol. Exp. Neurol.* **2002**. [CrossRef]
26. Paillusson, S.; Gomez-Suaga, P.; Stoica, R.; Little, D.; Gissen, P.; Devine, M.J.; Noble, W.; Hanger, D.P.; Miller, C.C.J. α-Synuclein binds to the ER–mitochondria tethering protein VAPB to disrupt Ca2+ homeostasis and mitochondrial ATP production. *Acta Neuropathol.* **2017**, *134*, 129–149. [CrossRef]
27. Bonello, F.; Hassoun, S.M.; Mouton-Liger, F.; Shin, Y.S.; Muscat, A.; Tesson, C.; Lesage, S.; Beart, P.M.; Brice, A.; Krupp, J.; et al. LRRK2 impairs PINK1/Parkin-dependent mitophagy via its kinase activity: Pathologic insights into Parkinson's disease. *Hum. Mol. Genet.* **2019**, *28*, 1645–1660. [CrossRef] [PubMed]
28. Shimura, H.; Hattori, N.; Kubo, S.I.; Mizuno, Y.; Asakawa, S.; Minoshima, S.; Shimizu, N.; Iwai, K.; Chiba, T.; Tanaka, K.; et al. Familial Parkinson disease gene product, parkin, is a ubiquitin-protein ligase. *Nat. Genet.* **2000**, *25*, 302–305. [CrossRef]
29. Dolgacheva, L.P.; Berezhnov, A.V.; Fedotova, E.I.; Zinchenko, V.P.; Abramov, A.Y. Role of DJ-1 in the mechanism of pathogenesis of Parkinson's disease. *J. Bioenerg. Biomembr.* **2019**, *51*, 175–188. [CrossRef] [PubMed]
30. Aerts, L.; de Strooper, B.; Morais, V.A. PINK1 activation—Turning on a promiscuous kinase. *Biochem. Soc. Trans.* **2015**. [CrossRef] [PubMed]
31. Jin, S.M.; Lazarou, M.; Wang, C.; Kane, L.A.; Narendra, D.P.; Youle, R.J. Mitochondrial membrane potential regulates PINK1 import and proteolytic destabilization by PARL. *J. Cell Biol.* **2010**, *191*, 933–942. [CrossRef] [PubMed]
32. Greene, A.W.; Grenier, K.; Aguileta, M.A.; Muise, S.; Farazifard, R.; Haque, M.E.; McBride, H.M.; Park, D.S.; Fon, E.A. Mitochondrial processing peptidase regulates PINK1 processing, import and Parkin recruitment. *EMBO Rep.* **2012**. [CrossRef] [PubMed]
33. Yamano, K.; Youle, R.J. PINK1 is degraded through the N-end rule pathway. *Autophagy* **2013**. [CrossRef] [PubMed]
34. Morais, V.A.; Verstreken, P.; Roethig, A.; Smet, J.; Snellinx, A.; Vanbrabant, M.; Haddad, D.; Frezza, C.; Mandemakers, W.; Vogt-Weisenhorn, D.; et al. Parkinson's disease mutations in PINK1 result in decreased complex I activity and deficient synaptic function. *EMBO Mol. Med.* **2009**. [CrossRef] [PubMed]
35. Pridgeon, J.W.; Olzmann, J.A.; Chin, L.S.; Li, L. PINK1 protects against oxidative stress by phosphorylating mitochondrial chaperone TRAP1. *PLoS Biol.* **2007**, *5*, 1494–1503. [CrossRef]
36. Plun-Favreau, H.; Klupsch, K.; Moisoi, N.; Gandhi, S.; Kjaer, S.; Frith, D.; Harvey, K.; Deas, E.; Harvey, R.J.; McDonald, N.; et al. The mitochondrial protease HtrA2 is regulated by Parkinson's disease-associated kinase PINK1. *Nat. Cell Biol.* **2007**, *9*, 1243–1252. [CrossRef]
37. Arena, G.; Gelmetti, V.; Torosantucci, L.; Vignone, D.; Lamorte, G.; de Rosa, P.; Cilia, E.; Jonas, E.A.; Valente, E.M. PINK1 protects against cell death induced by mitochondrial depolarization, by phosphorylating Bcl-xL and impairing its pro-apoptotic cleavage. *Cell Death Differ.* **2013**, *20*, 920–930. [CrossRef]
38. Morais, V.A.; Haddad, D.; Craessaerts, K.; de Bock, P.J.; Swerts, J.; Vilain, S.; Aerts, L.; Overbergh, L.; Grünewald, A.; Seibler, P.; et al. PINK1 loss-of-function mutations affect mitochondrial complex I activity via NdufA10 ubiquinone uncoupling. *Science* **2014**, *344*, 203–207. [CrossRef] [PubMed]
39. Okatsu, K.; Oka, T.; Iguchi, M.; Imamura, K.; Kosako, H.; Tani, N.; Kimura, M.; Go, E.; Koyano, F.; Funayama, M.; et al. PINK1 autophosphorylation upon membrane potential dissipation is essential for Parkin recruitment to damaged mitochondria. *Nat. Commun.* **2012**, *3*. [CrossRef]
40. Lee, J.Y.; Nagano, Y.; Taylor, J.P.; Lim, K.L.; Yao, T.P. Disease-causing mutations in Parkin impair mitochondrial ubiquitination, aggregation, and HDAC6-dependent mitophagy. *J. Cell Biol.* **2010**, *189*, 671–679. [CrossRef]
41. Aerts, L.; Craessaerts, K.; de Strooper, B.; Morais, V.A. PINK1 kinase catalytic activity is regulated by phosphorylation on serines 228 and 402. *J. Biol. Chem.* **2015**, *290*, 2798–2811. [CrossRef]
42. Kondapalli, C.; Kazlauskaite, A.; Zhang, N.; Woodroof, H.I.; Campbell, D.G.; Gourlay, R.; Burchell, L.; Walden, H.; MacArtney, T.J.; Deak, M.; et al. PINK1 is activated by mitochondrial membrane potential depolarization and stimulates Parkin E3 ligase activity by phosphorylating Serine 65. *Open Biol.* **2012**. [CrossRef]
43. McLelland, G.L.; Soubannier, V.; Chen, C.X.; McBride, H.M.; Fon, E.A. Parkin and PINK1 function in a vesicular trafficking pathway regulating mitochondrial quality control. *EMBO J.* **2014**, *33*, 282–295. [CrossRef]
44. Soubannier, V.; Rippstein, P.; Kaufman, B.A.; Shoubridge, E.A.; McBride, H.M. Reconstitution of mitochondria derived vesicle formation demonstrates selective enrichment of oxidized cargo. *PLoS ONE* **2012**, *7*, e52830. [CrossRef]

45. Braak, H.; Rüb, U.; Gai, W.P.; del Tredici, K. Idiopathic Parkinson's disease: Possible routes by which vulnerable neuronal types may be subject to neuroinvasion by an unknown pathogen. *J. Neural. Transm.* **2003**, *110*, 517–536. [CrossRef] [PubMed]
46. Braak, H.; Braak, E. Pathoanatomy of Parkinson's disease. *J. Neurol.* **2000**. [CrossRef] [PubMed]
47. Wang, D.; Qian, L.; Xiong, H.; Liu, J.; Neckameyer, W.S.; Oldham, S.; Xia, K.; Wang, J.; Bodmer, R.; Zhang, Z. Antioxidants protect PINK1-dependent dopaminergic neurons in drosophila. *Proc. Natl. Acad. Sci. USA* **2006**, *103*, 13520–13525. [CrossRef] [PubMed]
48. Valente, E.M.; Abou-Sleiman, P.M.; Caputo, V.; Muqit, M.M.K.; Harvey, K.; Gispert, S.; Ali, Z.; del Turco, D.; Bentivoglio, A.R.; Healy, D.G.; et al. Hereditary early-onset Parkinson's disease caused by mutations in PINK1. *Science* **2004**, *304*, 1158–1160. [CrossRef]
49. Valente, E.M.; Salvi, S.; Ialongo, T.; Marongiu, R.; Elia, A.E.; Caputo, V.; Romito, L.; Albanese, A.; Dallapiccola, B.; Bentivoglio, A.R. PINK1 mutations are associated with sporadic early-onset Parkinsonism. *Ann. Neurol.* **2004**, *56*, 336–341. [CrossRef]
50. Liu, X.L.; di Wang, Y.; Yu, X.M.; Li, D.W.; Li, G.R. Mitochondria-mediated damage to dopaminergic neurons in Parkinson's disease (review). *Int. J. Mol. Med.* **2018**, *41*, 615–623. [CrossRef]
51. Birsa, N.; Norkett, R.; Higgs, N.; Lopez-Domenech, G.; Kittler, J.T. Mitochondrial trafficking in neurons and the role of the Miro family of GTPase proteins. *Biochem. Soc. Trans.* **2013**, *41*, 1525–1531. [CrossRef]
52. Wang, X.; Winter, D.; Ashrafi, G.; Schlehe, J.; Wong, Y.L.; Selkoe, D.; Rice, S.; Steen, J.; Lavoie, M.J.; Schwarz, T.L. PINK1 and Parkin target miro for phosphorylation and degradation to arrest mitochondrial motility. *Cell* **2011**. [CrossRef]
53. Chen, Y.; Dorn, G.W. PINK1-phosphorylated mitofusin 2 is a parkin receptor for culling damaged mitochondria. *Science* **2013**, *340*, 471–475. [CrossRef] [PubMed]
54. Pryde, K.R.; Smith, H.L.; Chau, K.Y.; Schapira, A.H.V. PINK1 disables the anti-fission machinery to segregate damaged mitochondria for mitophagy. *J. Cell Biol.* **2016**, *213*, 163–171. [CrossRef]
55. Otera, H.; Mihara, K. Molecular mechanisms and physiologic functions of mitochondrial dynamics. *J. Biochem.* **2011**, *149*, 241–251. [CrossRef] [PubMed]
56. Han, H.; Tan, J.; Wang, R.; Wan, H.; He, Y.; Yan, X.; Guo, J.; Gao, Q.; Li, J.; Shang, S.; et al. PINK 1 phosphorylates Drp1 S616 to regulate mitophagy-independent mitochondrial dynamics. *EMBO Rep.* **2020**, *21*. [CrossRef] [PubMed]
57. Wood-Kaczmar, A.; Gandhi, S.; Yao, Z.; Abramov, A.S.Y.; Miljan, E.A.; Keen, G.; Stanyer, L.; Hargreaves, I.; Klupsch, K.; Deas, E.; et al. PINK1 is necessary for long term survival and mitochondrial function in human dopaminergic neurons. *PLoS ONE* **2008**, *3*, e2455. [CrossRef]
58. Pirooznia, S.K.; Yuan, C.; Khan, M.R.; Karuppagounder, S.S.; Wang, L.; Xiong, Y.; Kang, S.U.; Lee, Y.; Dawson, V.L.; Dawson, T.M. PARIS induced defects in mitochondrial biogenesis drive dopamine neuron loss under conditions of parkin or PINK1 deficiency. *Mol. Neurodegener.* **2020**, *15*. [CrossRef] [PubMed]
59. Gandhi, S.; Wood-Kaczmar, A.; Yao, Z.; Plun-Favreau, H.; Deas, E.; Klupsch, K.; Downward, J.; Latchman, D.S.; Tabrizi, S.J.; Wood, N.W.; et al. PINK1-associated Parkinson's disease is caused by neuronal vulnerability to calcium-induced cell death. *Mol. Cell* **2009**. [CrossRef]
60. Gandhi, S.; Muqit, M.M.K.; Stanyer, L.; Healy, D.G.; Abou-Sleiman, P.M.; Hargreaves, I.; Heales, S.; Ganguly, M.; Parsons, L.; Lees, A.J.; et al. PINK1 protein in normal human brain and Parkinson's disease. *Brain* **2006**, *129*, 1720–1731. [CrossRef] [PubMed]
61. Lin, Q.; Li, S.; Jiang, N.; Shao, X.; Zhang, M.; Jin, H.; Zhang, Z.; Shen, J.; Zhou, Y.; Zhou, W.; et al. PINK1-parkin pathway of mitophagy protects against contrast-induced acute kidney injury via decreasing mitochondrial ROS and NLRP3 inflammasome activation. *Redox Biol.* **2019**. [CrossRef]
62. Li, X.; Shi, Z.; Zhu, Y.; Shen, T.; Wang, H.; Shui, G.; Loor, J.J.; Fang, Z.; Chen, M.; Wang, X.; et al. Cyanidin-3-O-glucoside improves non-alcoholic fatty liver disease by promoting PINK1-mediated mitophagy in mice. *Br. J. Pharmacol.* **2020**. [CrossRef]
63. Wang, B.; Nie, J.; Wu, L.; Hu, Y.; Wen, Z.; Dong, L.; Zou, M.H.; Chen, C.; Wang, D.W. AMPKa2 protects against the development of heart failure by enhancing mitophagy via PINK1 phosphorylation. *Circ. Res.* **2018**. [CrossRef]
64. Matheoud, D.; Cannon, T.; Voisin, A.; Penttinen, A.M.; Ramet, L.; Fahmy, A.M.; Ducrot, C.; Laplante, A.; Bourque, M.J.; Zhu, L.; et al. Intestinal infection triggers Parkinson's disease-like symptoms in Pink1$^{-/-}$ mice. *Nature* **2019**, *571*, 565–569. [CrossRef] [PubMed]
65. Benarroch, E.E. Astrocyte signaling and synaptic homeostasis II: Astrocyte-neuron interactions and clinical correlations. *Neurology* **2016**, *87*, 726–735. [CrossRef] [PubMed]
66. Herculano-Houzel, S. The human brain in numbers: A linearly scaled-up primate brain. *Front. Hum. Neurosci.* **2009**, *3*, 31. [CrossRef] [PubMed]
67. Lin, L.F.H.; Doherty, D.H.; Lile, J.D.; Bektesh, S.; Collins, F. GDNF: A glial cell line—Derived neurotrophic factor for midbrain dopaminergic neurons. *Science* **1993**, *260*, 1130–1132. [CrossRef]
68. Kwon, H.S.; Koh, S.H. Neuroinflammation in neurodegenerative disorders: The roles of microglia and astrocytes. *Transl. Neurodegener.* **2020**, *9*, 1–12. [CrossRef] [PubMed]
69. Miklossy, J.; Doudet, D.D.; Schwab, C.; Yu, S.; McGeer, E.G.; McGeer, P.L. Role of ICAM-1 in persisting inflammation in Parkinson disease and MPTP monkeys. *Exp. Neurol.* **2006**, *197*, 275–283. [CrossRef]
70. Koprich, J.B.; Reske-Nielsen, C.; Mithal, P.; Isacson, O. Neuroinflammation mediated by IL-1β increases susceptibility of dopamine neurons to degeneration in an animal model of Parkinson's disease. *J. Neuroinflamm.* **2008**, *5*. [CrossRef]

71. Di Domenico, A.; Carola, G.; Calatayud, C.; Pons-Espinal, M.; Muñoz, J.P.; Richaud-Patin, Y.; Fernandez-Carasa, I.; Gut, M. Faella, A.; Parameswaran, J.; et al. Patient-specific iPSC-derived astrocytes contribute to non-cell-autonomous neurodegeneration in Parkinson's disease. *Stem Cell Rep.* **2019**, *12*, 213–229. [CrossRef]
72. Lindström, V.; Gustafsson, G.; Sanders, L.H.; Howlett, E.H.; Sigvardson, J.; Kasrayan, A.; Ingelsson, M.; Bergström, J.; Erlandsson A. Extensive uptake of α-synuclein oligomers in astrocytes results in sustained intracellular deposits and mitochondrial damage *Mol. Cell. Neurosci.* **2017**, *82*, 143–156. [CrossRef]
73. Wakida, N.M.; Cruz, G.M.S.; Ro, C.C.; Moncada, E.G.; Khatibzadeh, N.; Flanagan, L.A.; Berns, M.W. Phagocytic response of astrocytes to damaged neighboring cells. *PLoS ONE* **2018**, *13*, e0196153. [CrossRef]
74. Morales, I.; Sanchez, A.; Puertas-Avendaño, R.; Rodriguez-Sabate, C.; Perez-Barreto, A.; Rodriguez, M. Neuroglial transmitophagy and Parkinson's disease. *Glia* **2020**, *68*, 2277–2299. [CrossRef]
75. English, K.; Shepherd, A.; Uzor, N.E.; Trinh, R.; Kavelaars, A.; Heijnen, C.J. Astrocytes rescue neuronal health after cisplatin treatment through mitochondrial transfer. *Acta Neuropathol. Commun.* **2020**, *8*. [CrossRef]
76. Choi, I.; Choi, D.J.; Yang, H.; Woo, J.H.; Chang, M.Y.; Kim, J.Y.; Sun, W.; Park, S.M.; Jou, I.; Lee, S.H.; et al. PINK1 expression increases during brain development and stem cell differentiation, and affects the development of GFAP-positive astrocytes. *Mol Brain* **2016**, *9*. [CrossRef]
77. Barodia, S.K.; McMeekin, L.J.; Creed, R.B.; Quinones, E.K.; Cowell, R.M.; Goldberg, M.S. PINK1 phosphorylates ubiquitin predominantly in astrocytes. *NPJ Park. Dis.* **2019**, *5*. [CrossRef]
78. Choi, I.; Kim, J.; Jeong, H.K.; Kim, B.; Jou, I.; Park, M.; Chen, L.; Kang, U.J.; Zhuang, X.; Joe, E. Hye PINK1 deficiency attenuates astrocyte proliferation through mitochondrial dysfunction, reduced AKT and increased p38 MAPK activation, and downregulation of EGFR. *Glia* **2013**, *61*, 800–812. [CrossRef] [PubMed]
79. Sun, L.; Shen, R.; Agnihotri, S.K.; Chen, Y.; Huang, Z.; Büeler, H. Lack of PINK1 alters glia innate immune responses and enhances inflammation-induced, nitric oxide-mediated neuron death. *Sci. Rep.* **2018**, *8*. [CrossRef]
80. Liddelow, S.A.; Guttenplan, K.A.; Clarke, L.E.; Bennett, F.C.; Bohlen, C.J.; Schirmer, L.; Bennett, M.L.; Münch, A.E.; Chung, W.S. Peterson, T.C.; et al. Neurotoxic reactive astrocytes are induced by activated microglia. *Nature* **2017**, *541*, 481–487. [CrossRef] [PubMed]
81. Sofroniew, M.V. Molecular dissection of reactive astrogliosis and glial scar formation. *Trends Neurosci.* **2009**, *32*, 638–647. [CrossRef] [PubMed]
82. Ho, M.S. Microglia in Parkinson's disease. In *Advances in Experimental Medicine and Biology*; Springer: Berlin/Heidelberg, Germany 2019; Volume 1175, pp. 335–353.
83. Du, L.; Zhang, Y.; Chen, Y.; Zhu, J.; Yang, Y.; Zhang, H.L. Role of microglia in neurological disorders and their potentials as a therapeutic target. *Mol. Neurobiol.* **2016**, *54*, 7567–7584. [CrossRef] [PubMed]
84. Napoli, I.; Neumann, H. Protective effects of microglia in multiple sclerosis. *Exp. Neurol.* **2010**, *225*, 24–28. [CrossRef]
85. Zhang, W.; Wang, T.; Pei, Z.; Miller, D.S.; Wu, X.; Block, M.L.; Wilson, B.; Zhang, W.; Zhou, Y.; Hong, J.-S.; et al. Aggregated α-synuclein activates microglia: A process leading to disease progression in Parkinson's disease. *FASEB J.* **2005**, *19*, 533–542 [CrossRef] [PubMed]
86. Pósfai, B.; Cserép, C.; Orsolits, B.; Dénes, Á. New insights into microglia-neuron interactions: A neuron's perspective. *Neuroscience* **2019**, *405*, 103–117. [CrossRef] [PubMed]
87. Tremblay, M.E.; Cookson, M.R.; Civiero, L. Glial phagocytic clearance in Parkinson's disease. *Mol. Neurodegener.* **2019**, *14*, 1–14 [CrossRef]
88. Gerhard, A.; Pavese, N.; Hotton, G.; Turkheimer, F.; Es, M.; Hammers, A.; Eggert, K.; Oertel, W.; Banati, R.B.; Brooks, D.J. In vivo imaging of microglial activation with [11C](R)-PK11195 PET in idiopathic Parkinson's disease. *Neurobiol. Dis.* **2006**, *21*, 404–412 [CrossRef]
89. Yun, S.P.; Kam, T.I.; Panicker, N.; Kim, S.; Oh, Y.; Park, J.S.; Kwon, S.H.; Park, Y.J.; Karuppagounder, S.S.; Park, H.; et al. Block of A1 astrocyte conversion by microglia is neuroprotective in models of Parkinson's disease. *Nat. Med.* **2018**, *24*, 931–938. [CrossRef] [PubMed]
90. Rojanathammanee, L.; Murphy, E.J.; Combs, C.K. Expression of mutant alpha-synuclein modulates microglial phenotype in vitro *J. Neuroinflamm.* **2011**, *8*. [CrossRef]
91. Kim, J.; Byun, J.-W.; Choi, I.; Kim, B.; Jeong, H.-K.; Jou, I.; Joe, E. PINK1 deficiency enhances inflammatory cytokine release from acutely prepared brain slices. *Exp. Neurobiol.* **2013**, *22*, 38–44. [CrossRef]
92. Matheoud, D.; Sugiura, A.; Bellemare-Pelletier, A.; Laplante, A.; Rondeau, C.; Chemali, M.; Fazel, A.; Bergeron, J.J.; Trudeau, L.E. Burelle, Y.; et al. Parkinson's disease-related proteins PINK1 and parkin repress mitochondrial antigen presentation. *Cell* **2016** *166*, 314–327. [CrossRef] [PubMed]
93. Deeb, W.; Nozile-Firth, K.; Okun, M.S. Parkinson's disease: Diagnosis and appreciation of comorbidities. In *Handbook of Clinical Neurology*; Elsevier: Amsterdam, The Netherlands, 2019; Volume 167, pp. 257–277.
94. Ntetsika, T.; Papathoma, P.E.; Markaki, I. Novel targeted therapies for Parkinson's disease. *Mol. Med.* **2021**, *27*, 1–20. [CrossRef]
95. George, S.; Rey, N.L.; Tyson, T.; Esquibel, C.; Meyerdirk, L.; Schulz, E.; Pierce, S.; Burmeister, A.R.; Madaj, Z.; Steiner, J.A.; et al Microglia affect α-synuclein cell-to-cell transfer in a mouse model of Parkinson's disease. *Mol. Neurodegener.* **2019**, *14*, 1–22 [CrossRef] [PubMed]

Review

Alpha-Synuclein and Lipids: The Elephant in the Room?

Alessia Sarchione, Antoine Marchand, Jean-Marc Taymans and Marie-Christine Chartier-Harlin *

Univ. Lille, Inserm, CHU Lille, UMR-S 1172—LilNCog—Lille Neuroscience and Cognition, F-59000 Lille, France; Alessia.Sarchione@inserm.fr (A.S.); Antoine.Marchand@inserm.fr (A.M.); jean-marc.taymans@inserm.fr (J.-M.T.)
* Correspondence: marie-christine.chartier-harlin@inserm.fr

Abstract: Since the initial identification of alpha-synuclein (α-syn) at the synapse, numerous studies demonstrated that α-syn is a key player in the etiology of Parkinson's disease (PD) and other synucleinopathies. Recent advances underline interactions between α-syn and lipids that also participate in α-syn misfolding and aggregation. In addition, increasing evidence demonstrates that α-syn plays a major role in different steps of synaptic exocytosis. Thus, we reviewed literature showing (1) the interplay among α-syn, lipids, and lipid membranes; (2) advances of α-syn synaptic functions in exocytosis. These data underscore a fundamental role of α-syn/lipid interplay that also contributes to synaptic defects in PD. The importance of lipids in PD is further highlighted by data showing the impact of α-syn on lipid metabolism, modulation of α-syn levels by lipids, as well as the identification of genetic determinants involved in lipid homeostasis associated with α-syn pathologies. While questions still remain, these recent developments open the way to new therapeutic strategies for PD and related disorders including some based on modulating synaptic functions.

Keywords: α-synuclein; exocytosis; genetics; lipids; membranes; Parkinson disease; SNARE complex; synapse; vesicle fusion; therapeutic target

Citation: Sarchione, A.; Marchand, A.; Taymans, J.-M.; Chartier-Harlin, M.-C. Alpha-Synuclein and Lipids: The Elephant in the Room? *Cells* **2021**, *10*, 2452. https://doi.org/10.3390/cells10092452

Academic Editor: Cristine Alves Da Costa

Received: 19 July 2021
Accepted: 12 September 2021
Published: 17 September 2021

1. Introduction

Parkinson's disease (PD) is one of the main neurodegenerative disorders, whose development is mainly due to the combined result of environmental factors and genetic predispositions, and based on the age at which symptoms appear, can be classified as juvenile, early onset, or late onset [1]. The neurodegeneration mainly affects the survival of dopamine producing neurons of the substantia nigra pars compacta, and both the premature degeneration of dopaminergic neurons and accumulation of protein-rich aggregates, called Lewy bodies, are the main neuropathological hallmarks of PD [2]. Post-mortem diagnosis of pre-symptomatic stages of the disease is based on the identification of these inclusion bodies, which develop as spindle-like Lewy neurites in cellular processes and as globular Lewy bodies in neuronal cell bodies [3]. These hallmarks are associated with consistent activation of microglia surrounding degenerating dopaminergic neurons in the substantia nigra, suggesting an important role of the immune system in this disorder [4]. At present, no curative treatments for PD are available, putting forward the need to better understand the mechanisms leading to the neurodegeneration of the nigrostriatal system. This might come from a better understanding of the role of a key protein involved in this disorder, namely alpha-synuclein (α-syn).

The α-syn protein is encoded by the *Non-A4 Component Of Amyloid Precursor* (*SNCA*) gene that is located at the PARK1/4 locus on chromosome 4q21 and consists of six protein coding exons [5–7]. While PD is mainly sporadic, several deleterious or potentially deleterious mutations in this gene (A18T, A29S, A30G, A30P, E46K, H50Q, G51D, A53E, A53T, and A53V) have been linked to familial parkinsonism [8–11] (Figure 1a). Further evidence, including triplication [12] and duplication of the *SNCA* gene locus [13,14], demonstrates that the sole overexpression of α-syn can lead to the disease. Families with *SNCA* mutations or locus multiplications are relatively rare; however, several case control studies and genome-wide association studies (GWAS) demonstrated that polymorphisms at this gene locus also are moderate risk factors for PD [15–17]. Furthermore, post-transcriptional effects on *SNCA* transcripts, such as usage of alternative start sites and variable UTR lengths exist [18,19], leading to more than 40 transcripts, at least some of which are associated to PD [20]. Epigenetic deregulation in the *SNCA* gene is also associated with idiopathic PD [21]. In addition, *SNCA* copy number variant mosaicism has been reported [22–24]. Further studies are needed to confirm the roles of transcript, epigenetic, and mosaicism variants in the pathogenesis of PD. Overall, the *SNCA* gene is one of the most important genetic determinants involved in the pathogenesis of PD [25,26].

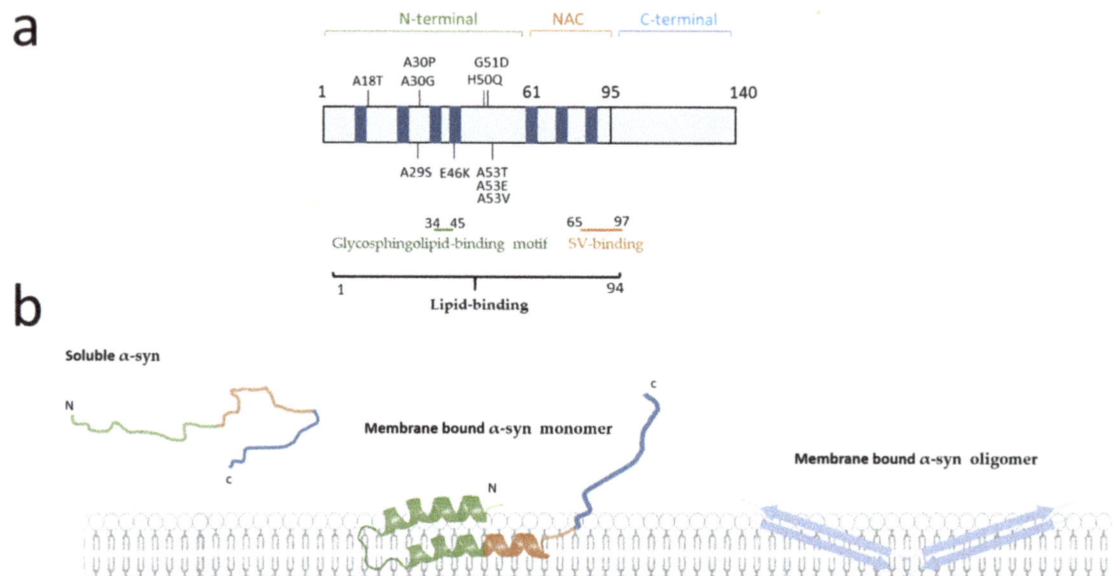

Figure 1. Schematic representation of α-synuclein (α-syn) mutations and lipid binding regions. (**a**) Schematic representation of the domain structure of α-syn. The α-syn is composed of three domains: the N-terminal domain (green), the NAC domain (orange) and the C-terminal domain (blue). Four confirmed pathogenic autosomal dominant missense mutations (A30P, G51D, A53E, A53T) as well as six putatively pathogenic mutations (A18T, A29S, A30G, E46K, H50Q, A53V) are depicted [11]. In blue are represented the seven KTKEGV hexameric repeats spanning from the N-terminus to the non-amyloid β-component (NAC) domain. The lipid binding regions are represented by lines of different colours (black = lipid binding, green = glycosphingolipid-binding motif and red = synaptic vesicles (SV)). (**b**) Schematic representation of the different conformations of the α-syn. α-syn is present in the cytosol as unfolded monomer. Binding of α-syn to lipids induces a conformational change of α-syn N-terminal region, which acquires an α-helix secondary structure. The oligomers penetrate into the lipid bilayer with a β-sheet structure. The membrane image is adapted from Servier Medical Art (smart.servier.com, accessed on 19 July 2020) licensed under a Creative Commons Attribution 3.0 Unported License.

Additional arguments point to the major role of α-syn in neurodegenerative disorders. Indeed, it has long been established that aggregated α-syn is a hallmark of synucleinopathies, including the presence of α-syn positive Lewy bodies in the neurons of PD, dementia with Lewy bodies [27] and some variants of Alzheimer's disease [27,28]. In addition, aggregated α-syn has been observed in glial cells in multiple system atrophy [29]. Spontaneous conversion of soluble unfolded α-syn monomers into aggregates leads to accumulation of α-syn in neurons. The most common form of α-syn is thought to be monomeric and found in the cytoplasm of neuronal cell models [30], whereas under pathological conditions α-syn is thought to form oligomers (Figure 1b). Intriguingly, under physiological conditions, α-syn is able to form helically folded tetramers that might be more resistant to aggregation. However, these data need still to be better understood [31]. Conversely, the spread of insoluble α-syn propagation from cell-to-cell is currently considered as a mechanism to explain the pathological progression of disease along synaptically connected regions of the brain [32,33]. Furthermore, many studies in post-mortem brains, indicate that the degree of microglial activation in PD is directly correlated with α-syn deposition, suggesting that α-syn may be directly involved in activating the innate immune system [4]. Similarly, recent data have shown overexpression of α-syn in human induced pluripotent stem cells (iPSC) derived neurons and in neuronal tissues of non-human primates after viral infection, further bolstering the hypothesised link between immune system challenge and synucleinopathies [34].

In addition, the α-syn protein is involved in a wide range of processes impaired in PD pathophysiology including transport of synaptic vesicles (SV), regulation of dopamine release, and vesicular trafficking. Indeed, α-syn physiologically interacts with membrane lipids (Figure 1b) and proteins in order to regulate synaptic plasticity and neurotransmitter release [35]. A current hypothesis is that α-syn dysfunction can lead to defects in vesicular trafficking and several studies conducted in worm, yeast, fly, and mouse models tend to confirm this assumption [36]. Further evidence supporting the ability of α-syn to regulate membrane trafficking processes is directly correlated with its interaction with membrane lipids and several proteins, especially at the synapse. Among the partners of α-syn, a crucial role has emerged for instance for SNARE proteins (soluble N-ethylmaleimide-sensitive factor (NSF) attachment protein (SNAP) receptors), which represent the core machinery mediating vesicle trafficking and membrane fusion. The orchestrated coordination of α-syn and SNARE proteins allows the regulation of synaptic plasticity and neurotransmitter release [35,37]. Interestingly, an emerging dimension to the role of α-syn in membrane trafficking is the importance of membrane lipid composition, with recent evidence showing for example that membrane lipid composition modulates the role of α-syn in neurotransmitter release [38]. Thus, we aim to examine in a first step the physical relationship between α-syn and lipids in the context of plasma and SV membranes. Secondly, we will describe the implications of these interactions on synaptic functions of α-syn, including docking, exocytosis, and recycling of SV. The final goal is to discuss the lipid deregulations in PD and potential therapeutic strategies for synucleinopathies.

2. α-Synuclein and Its Relationship with Lipid Membranes

α-syn was originally described as a protein enriched at the synapse [39] and was later identified as a component of Lewy bodies in PD [40]. Of particular importance recently, we learned that these inclusions are also enriched in lipid membranes and degenerated organelles [41]. These data first suggested a role for α-syn at the synapse and recent advances on the composition of Lewy bodies highlight a strong relationship between α-syn and membranes as well as lipids. Moreover, the role of α-syn in synaptic activity implies the need to decipher the mechanism of interaction of α-syn with biological membranes.

2.1. α-Synuclein Structure and Interaction with Lipids

Biophysical studies reveal that α-syn interacts with lipid components of biological membranes in different manners. The specific nature, affinity, and functional effects of these interactions have been extensively investigated by in vitro studies performed on artificial membrane systems of different levels of complexity (summarised in Table 1).

Table 1. Presentation of membrane models used to investigate the basic physical and biochemical role of α-synuclein (α-syn). The artificial membrane systems are used to study the physical interaction of α-syn and lipids. They are classified into two categories according to their three-dimensional organisation: vesicular and planar models. These systems can be created using different types and proportions of phospholipids allowing the study of different binding properties of α-syn.

Membrane Model	Description	Principal Fields of Investigation
Vesicular systems		
Micelles	Spherical and monolayer system of amphipathic molecules. Substantial difference with biological membranes.	To identify the conformational change of α-syn domains upon interaction with lipids [42].
Liposomes	Spherical vesicles composed of at least one lipid bilayer and of different sizes and curvatures [43] (1) SUV* of 10–100 nm; (2) LUV* 100 nm; (3) GUV* 1 μm.	To investigate the effect of membrane curvature on α-syn oligomer–membrane interactions based on the size: (1) SUV interaction of α-syn with SV; (2) LUV mimicking cell membrane organelles; (3) GUV α-syn relationship with cell membrane [43].
Planar systems		
Lipid monolayer or bilayer	Planar structure composed of one or two layers.	To investigate the interaction between oligomers and membranes and to analyse the effect of α-syn oligomers on membrane disruption [43].
Nanodisc	Planar bilayer structure composed of (1) phospholipids of artificial or cell membrane origin. (2) scaffolding proteins or polymers conferring stability to the system. Size variability from 7 to 50 nm. High similarity to biological membranes.	To allow structuring of disordered proteins, such as α-syn into non-toxic α-helical structures [44].

Legend. α-syn = α-synuclein, GUV* = giant unilamellar vesicles, LUV* = large unilamellar vesicles, SUV* = small unilamellar vesicles, SV = synaptic vesicles.

The studies in membrane-mimicking models investigate the interaction between different classes of lipids and the three α-syn domains: the positively charged N-terminal domain (residues 1–60), a central hydrophobic NAC (non-amyloid β-component) domain (residues 61–95), and the acidic C-terminal tail (residues 96–140). The different domains and motifs of α-syn are schematically depicted in Figure 1. The basic character of the N-terminal domain allows the formation of electrostatic interactions with acidic negatively charged membrane lipids [45] particularly enriched in the membrane of SV [46]. The N-terminal domain shows an affinity for glycosphingolipids and, specifically, the residues 34–45 have been proposed as a cell surface lipid-binding motif bearing a solvent-accessible aromatic residue [47]. It should be noted that such a domain is also present on other proteins responsible for neurodegeneration such as prion protein and amyloid β [48]. The binding of the N-terminal domain of α-syn to lipids induces a conformational change from a random-coil to a more stable α-helix structure [49].

The α-syn protein sequence has several characteristic imperfect repeats of 11 amino acids extending from the N-terminus to the NAC domain with a highly conserved hexameric sequence (KTKEGV), which is also present in the α-helix motif of the lipid-binding domains of apolipoproteins A2 [42]. These repeats have the propensity to adopt an α-helical structure upon binding with negatively charged phospholipid membranes. Studies on sodium dodecyl sulphate-micelles suggest that α-syn-micelle bonds involve a long α-helical region (from residue 1–94) interrupted by a short linker including residues 42, 43 and 44. These data are in contrast to previous evidence from Davidson et al. showing the existence of five α-helices of α-syn bound to liposomes [50]. The two models are not considered mutually exclusive and the switch between the two conformations depends on membrane lipid rearrangement and organisation [51]. The central NAC domain is the most hydrophobic part of α-syn and is prone to acquire a β-sheet conformation [52]. It represents the domain leading to the nucleation of α-syn in oligomer formation. The NAC region might be partially inserted into the lipid bilayer [53], but its most important role is to act as a modulator of α-syn affinity for lipid membranes [54].

The C-terminal domain, enriched in proline residues, is an unstructured region likely due to its low hydrophobicity and confers flexibility to the protein. The C-terminus is weakly associated with the membrane [54], but it has recently been shown that calcium increases the membrane association of this domain. The random coil configuration of the acidic carboxylic tail is conserved also in the α-syn lipid bound state [55]. In addition, this α-syn domain undergoes several post-translational modifications, the best known being the S129 phosphorylation that accumulates within Lewy bodies [56].

2.2. α-Synuclein and Lipid Bilayers

Biological membranes exhibit a heterogeneity in lipid composition as well as asymmetry in the proportions and distribution of lipids between the two leaflets of the lipid bilayer. This asymmetric lipid composition will influence the binding affinity of α-syn to the presynaptic and SV membranes.

2.2.1. Presynaptic Membrane Composition and α-Synuclein Binding Affinity

Biological membranes are mainly composed of three different types of lipids classified as phospholipids, glycolipids, and cholesterol [57]. Studies on the lipid composition of the plasma membrane (PM) reveal that, among the phospholipids, the most represented in membrane include phosphatidylcholine (PC), phosphatidylethanolamines (PE), sphingomyelin, and cholesterol. These classes of lipids are found in both leaflets of the membrane. Nevertheless, biochemical analyses revealed the asymmetric distribution of lipids between the two leaflets of the bilayer called inner PM (IPM on the cytosolic side) and outer PM (OPM on the extracellular side). Interestingly, under physiological conditions, phospho-L-serine (PS), phosphatidylinositol (PI) and phosphatidylinositol phosphates (PIPs) are more specifically present on the IPM. In contrast, gangliosides (GM) and cerebrosides are more specific to OPM [58]. Importantly, based on this differential distribution, the relationship of α-syn to the two leaflets was studied in a series of in vitro experiments by Man et al. using artificial membranes as models reflecting the same asymmetric distribution between the two leaflets of biological membranes [38]. The authors show that the binding of α-syn to either leaflet of the PM is quite different with α-syn having a strong affinity for IPM compared to OPM with the N-terminal region having the higher binding strength. This study supports the hypothesis of double-anchor mechanism whereby α-syn binds simultaneously to the IPM through its N-terminal region and to SV through a motif located in the NAC domain (residues 65–97) which has a weak affinity for IPM. Moreover, knowing that many classes of lipids are altered in neurodegenerative disorders (Table 2), Man et al. then investigated the α-syn binding affinity with IPM or OPM according to the enrichment or not of GM components (Figure 2a). Indeed, GM has emerged as an important factor in maintaining neuronal functions [59] and, moreover, GM concentration is altered in neurodegenerations with 22% reduction in brain GM content in men with PD, no differences in women with PD [60] and a 45% reduction in GM content observed in late stages of Alzheimer's disease. Assessment of the affinity of α-syn for OPM and IPM according to GM enrichment in both leaflets draws further attention to the role of GM on α-syn binding region. A six-fold increase interaction of the α-syn region 65–97 was observed in IPM-GM compared to IPM, while the N-terminal region kept the same strong affinity of binding for IPM-GM as for IPM. These results were confirmed also by the conformation analyses using chemical exchange saturation transfer experiments [38]. Similarly, α-syn shows stronger binding to OPM-GM than to OPM. In particular, the residues 1–35 of α-syn at the N-terminus show the higher affinity to OPM-GM, whereas both regions 36–98 and the C-terminal region 99–140 have low affinity or no binding, respectively [38].

Table 2. Overview of the main lipid classes altered in PD patients and models and their effect on α-synuclein (α-syn). This table provides some examples of different classes of lipids (first column) whose levels are altered in samples and biofluids from PD patients compared to controls (second column). We have also reported examples of enzymes associated with lipid metabolism whose activity is deregulated in PD as well as examples of genetic risk factors for PD associated with lipid catabolism. In some cases, these alterations may directly affect the properties and homeostasis of α-syn (third column).

Lipid Classes	Alterations in PD Patients	Effects on α-Syn
Phospholipids		
Phosphatidylcholine (PC)	Decreased PC (34:5, 36:5, and 38:5) in the frontal cortex of PD brains [63]. Decreased PC species with polyunsaturated 3, 4, and 36 carbon in visual cortex of PD [63]. Increased PC 44:6 and 44:5 and decreased PC 35:6 in the plasma of PD patients [63]. Deregulated PC pathway across transcriptome data derived from SN and putamen of PD patients versus controls [64]. Increased PC in CSF of PD patients [65].	POPC bilayer affects the α-syn aggregation [66].
Phosphatidylethanolamine (PE)	Reduced PE in early PD but not in advanced PD [67]. Deregulated PE pathway across transcriptome data derived from SN and putamen of PD patients versus controls [64].	Reduced levels of PE in the phosphatidylserine decarboxylase deletion mutant (*psd1*Δ) increase cytoplasmic α-syn inclusion and enhance toxicity in yeast [68].
Phosphatidylinositol (PI)	Decreased PI in rat and human cortical neurons overexpressing α-syn [63]. Deregulated PI pathway across transcriptome data derived from SN and putamen of PD patients versus controls [64].	Decreased PI species in yeast as well as rat or human cortical neurons overexpressing α-syn [63].
Phosphatidylserine (PS)	Increased PS with 36:1, 36:2 and 38:3 fatty acyl side chains in PD frontal cortex [69]. Deregulated PS pathway across transcriptome data derived from SN and putamen of PD patients versus controls [64].	Facilitation of SNARE complex formation and SNARE-dependent vesicles docking upon α-syn interaction with PS and v-SNARE [69]. Accelerated aggregation on POPS bilayers compared to POPC [66].
Sphingolipids		
Sphingomyelin	Reduced in PD anterior cingulate cortex compared to controls [70]. Deregulated sphingomyelin pathway across transcriptome data derived from SN and putamen of PD patients versus controls [64].	Increased α-syn transcript and protein levels upon cell treatment with exogenous sphingomyelin [71].
Gangliosides (GM)	Increased in lipid rafts [72]; 22% reduction in GM brain content in PD male patients, with no differences for PD female [60].	Hypothesised to be involved in both inhibition or enhancement of the α-syn aggregation kinetics [73]. Accelerate α-syn aggregation in presence of high GM1 and GM3 ganglioside concentration in exosomes [74].
Ceramides	Reduced total ceramides in PD anterior cingulate cortex compared to controls [70]. Increased in CSF of PD patients [72].	Increased α-syn toxicity as well as α-syn oligomers formation are linked to alteration in ceramide content [75].
Saturated fatty acids		
Stearic acid	Increased in lipid rafts [72]. Increased in rat treated with 6-hydroxydopamine (6-OHDA) [76].	Interaction with α-syn [77].
Palmitic Acid (PA)	Increased in lipid rafts [72].	Increased of α-syn expression levels in Thy1-α-syn mouse model after diet enriched in palmitic acid [78].
Palmitoleic Acid	Decreased in CSF of PD patients [72].	

Table 2. *Cont.*

Lipid Classes	Alterations in PD Patients	Effects on α-Syn
Unsaturated fatty acids		
α-linolenic acid	Decreased in CSF of PD patients [72].	Promoted formation of α-syn oligomers and α-syn induced cytotoxicity [79].
Oleic acid (OA)	Decreased in CFS of PD patients [72].	Increased in response to increase concentration of α-syn monomers [63]. Decreased by stearoyl-CoA desaturase (SCD) inhibition reduced α-syn toxicity [80].
Unsaturated fatty acids Omega-3		
Eicosapentaenoic acid (EPA)	Decreased EPA in lipid rafts [72].	
Docosanoic acid (DHA)	Decreased DHA in lipid rafts of DLB brain [81]. Increased amount of DHA (22:6) in PD and DLB brains [79].	Increased in α-syn oligomerisation in a DHA dose-dependent manner [79]. Increased accumulation of soluble and insoluble neuronal α-syn in A53T α-syn mice fed with an enriched DHA diet [82].
Other lipids		
Lipids with high solubility in aqueous solution and short hydrocarbon chains.	NI	Induced amyloid fibril formation of α-syn [83].
Enzyme associated to lipid metabolism		
Sphingomyelinase	Increased activity in PD brain and increased ceramide level [84]. Of note, acid sphingomyelinase (ASMase) encoded by *SMPD1* is responsible for the hydrolysis of sphingomyelin into ceramide and phosphorylcholine and a reduced ASMase enzymatic activity was associated with an earlier age at onset *SMPD1* variants in PD vs. controls. These genetic variants impair the traffic of acid-sphingomyelinase to the lysosomes [85].	Increased α-syn levels in HeLa and BE(2)-M17 dopaminergic cells in *SMPD1* KO and KD [85].
Sphingosine kinase I	Reduced SPHKs activity under oxidative stress evoked by MPP+ [84].	Induced of α-syn secretion and propagation upon SPHK inhibition [86].
Phospholipase D1 enzyme (PLD1)	Reduced activity and expression level of PLD1 observed in DLB post-mortem brains [87].	PLD1 prevents α-syn accumulation by autophagic flux activation [87].
Glucocerebrosidase (GBA)	Reduced GCase activity in the SN and hippocampus of iPD patients [88].	Misfolded GCase interacts with α-syn and induces α-syn accumulation and aggregation [89].
Cathepsins D and E	Increased activity of cathepsin D in *PRKN*-PD-derived fibroblasts [78] or in iPSC-derived dopaminergic neurons from N370S-GBA PD [90]. Increased activity of cathepsin E in blood and CSF from PD patients (See for review [91,92]).	α-syn is degraded by lipid-associated cathepsin D [93].
β-hexosaminidase	Decreased activity in blood and CSF from PD patients [91].	Increased β-hexosaminidase activity rescues the neurodegeneration induced by α-syn in dopaminergic neurons of the rodent SN [94].
β-galactosidase	Increased activity in blood and CSF from PD patients [91,92].	NI

Legend. α-syn = α-synuclein, CSF = cerebrospinal fluid, DLB = Lewy body dementia, GCase = glucocerebrosidase, iPD = idiopathic PD patients, iPSC = induced pluripotent stem cells, MPP = 1-methyl-4-phenylpyridinium, KD = knockdown, KO = knockout, NI = no information, POPC = 1-palmitoyl-2-oleoyl-*sn*-glycero-3-phosphocholine, POPS = 1-palmitoyl-2-oleoyl-sn-glycero-3-phosphoserine, *PRKN* = Parkin gene, SN = Substantia Nigra.

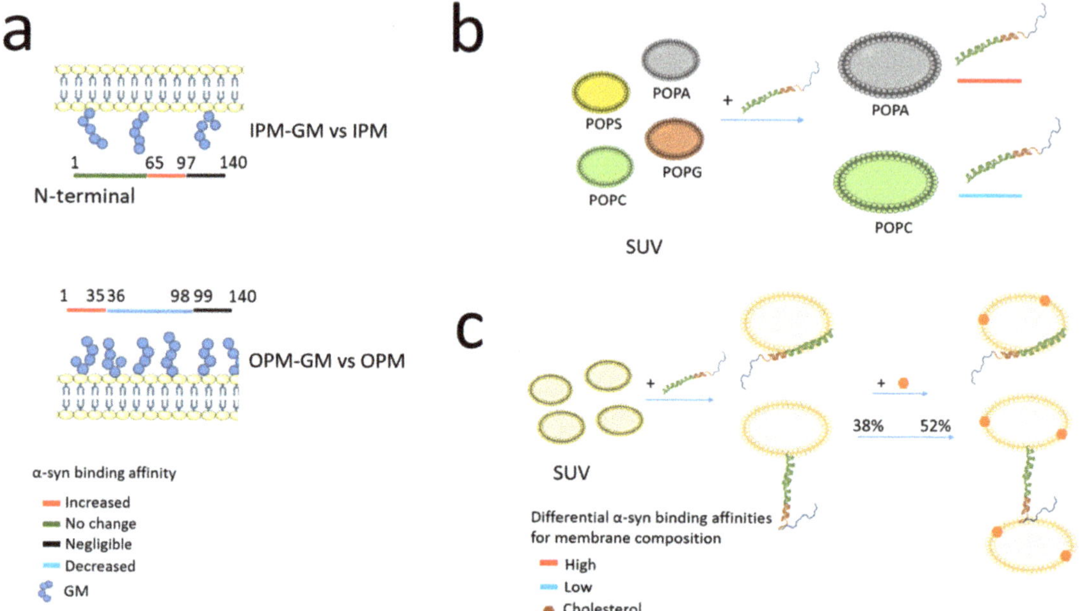

Figure 2. Schematic representation of differential affinity of α-synuclein (α-syn) for the inner or outer plasma membrane (IPM or OPM) as well as for vesicles according to their lipid compositions. (**a**) Differential affinity of α-syn for IPM and OPM according to differences in the amount of gangliosides (GM): IPM-GM versus (vs) IPM or OPM-GM versus OPM as described by Man et al. (2021) [38]. (**b**) Differential affinity of α-syn for artificial vesicles based on their membrane composition. α-syn has a 60 times higher affinity for 1-palmitoyl-2-oleoyl-*sn*-glycero-3-phosphate (POPA) than 1-palmitoyl-2-oleoyl-phosphatidyl-l-serine (POPS) and 1-palmitoyl-2-oleoyl-*sn*-glycero-3-phosphoglycerol (POPG) and very low affinity for 1-palmitoyl-2-oleoyl-sn-glycero-3-phosphocholine (POPC). (**c**) Effect of cholesterol on the conformation of α-syn. α-syn interacts with vesicles to promote fusion between 2 vesicles as described by Fusco et al. (2016) [61] and Man et al. (2020) [62]. Upon interaction with small unilamellar vesicles (SUV) composed of 1,2-dioleoyl-sn-glycero-3-phospho-ethanolamine (DOPE), 1,2-dioleoyl-sn-glycero-3-phospho-L-serine (DOPS), 1,2-dioleoyl-sn-glycero-3-phosphocholine (DOPC), α-syn exists in multiple different conformational states. These include the α-helical state covering the 1–97 region (top) and a conformational state interacting with the membrane through N-terminal residues 1–25. It has been proposed by Man et al. (2020) that the presence of cholesterol in the SUV composition induces an increase in the proportion of α-syn with the conformational state described at the bottom from 38% to 52%, leading to the 65–97 region being available to interact with a second SUV [62]. This suggests that cholesterol promotes the docking of the vesicles-mediated by α-syn.

These observations support the ability of α-syn to drive the docking of synaptic-like small unilamellar vesicles (SUV) to IPM in a concentration-dependent manner. Furthermore, if cholesterol levels are disturbed in PD patients, α-syn binding to OPM showing increased GM could be favoured. Therefore, the differential binding of α-syn to the two leaflets of the bilayer may have important implications in the synaptic activity of α-syn as described later in Section 3.1.

2.2.2. Lipid Rafts and α-Synuclein Interaction

On the PM, there are lipid microdomains called lipids rafts characterised by combinations of glycosphingolipids, cholesterol, and receptor proteins. Other lipids, such as relatively saturated phospholipids. have often been associated with raft-like environments [95]. They form functional platforms involved in the regulation of cellular functions and are present in both the inner and outer leaflets. The interaction of α-syn with lipid rafts is crucial in ensuring the synaptic localisation of α-syn. Indeed, knowing that in OPM glycosphingolipids are mostly present in sphingomyelin and cholesterol enriched lipid rafts, Fantini et al. determined the following ranking for the interaction of α-syn with glycosphingolipids [47]: GM3 > Gb3 > GalCer-NFA > GM1 > sulfatide > GalCer-HFA > LacCer > GM4 > GM2 > asialo-GM1 > GD3. Interestingly, the presence of GM3 stimulated the insertion of α-syn into sphingomyelin containing monolayers and promoted the integration of α-syn in raft-like membrane domains [47]. Furthermore, this association of α-syn with lipid rafts is dependent on ergosterol content and can be abrogated by depletion of cholesterol or by the presence of the α-syn A30P mutation. These two parameters also modify the preferential localisation of α-syn towards detergent-resistant fractions, corresponding to lipid raft domain of yeast membranes [96]. Note that Fortin et al. demonstrated in cellulo that synaptic localisation depends strongly on its interaction with the lipid rafts. Indeed, changes in lipid raft composition or affinity of α-syn in their binding may compromise the α-syn localisation and consequentially its normal function at the synapse in mouse brain [97]. Interestingly, Perissinotto et al. proposed another mechanism of preferential interaction, in which heavy metals play an important role in defining the lipid raft localisation of α-syn species [98]. In this study using atomic force microscopy, a thinning of the PM in the absence of ferrous cations Fe^{2+} and in the presence of monomers is observed. Knowing that heavy metal ions contribute to aggregations of monomers, the authors exposed the bilayer membrane model to Fe^{2+} and observed oligomer-like structures as expected. Interestingly, these aggregates were preferentially directed towards the lipid raft phase of the bilayer model [98]. In parallel, the authors show that the A53T mutated α-syn exhibited a greater and faster membrane interaction compared to wild type (WT) α-syn. If such models also exist in pathological conditions, this would further strengthen the role of lipids in PD pathophysiology.

2.3. *α-Synuclein and Synaptic Vesicles*

A large number of biophysical studies on α-syn and lipid interactions aimed to define the specificity and affinity of α-syn for synaptic-like vesicles as a function not only of lipid composition, but also of other parameters, including the size and curvature of vesicles.

2.3.1. α-Synuclein and Membrane Curvature

α-syn is capable of generating membrane curvature [99,100] and the synaptic concentration of α-syn is sufficient to induce membrane bending [101]. The curvature process occurs through the insertion of N-terminal region of α-syn into the membrane in a manner similar to other amphipathic helical proteins, such as endophilin [101]. Indeed, α-syn belongs to the class of proteins that can initiate a wedge in the bilayer (the amphipathic helices (9–41 AA)) and binds preferentially to pre-curved bilayers, where curvature has created a gap in lipid packing. Such a protein is considered as a curvature generator and curvature sensor [102]. Thanks to this ability, α-syn as well as other proteins such as β-syn and apolipoprotein A-1 are able to convert large vesicles into highly curved membrane tubules and vesicles [99]. However, compared to other curvature sensor proteins, α-syn does not use a bin/amphiphysin/rvs (BAR) domain and, therefore, has a lesser ability to induce tubulation compared to other proteins such as endophilin A1.

When studying the effect of different forms of α-syn, only monomeric, but not tetrameric, α-syn is able to induce membrane curvature. Moreover, the A30P mutant of α-syn, characterised by a distortion in its N-terminal domain and consequent disruption of α-helix formation, has a weak membrane binding, thus losing the ability to drive the membrane curvature [101]. In addition, the alterations in membrane trafficking observed in PD models of α-syn overexpression [103] were potentially associated with alterations in membrane curvature and membrane disruption induced by overexpression of α-syn [99]. Thereby, the membrane curvature mediated by α-syn may represent a crucial process allowing α-syn to fulfil a functional role in vesicle trafficking and vesicle exocytosis.

2.3.2. α-Synuclein Affinity According to Vesicle Composition

The lipid composition of vesicles deeply affects the binding, the state, and the solubility of α-syn, as documented above. Although physical interaction with lipid components of vesicles is crucial in the synaptic activity of α-syn, the affinity of α-syn for vesicles can change depending on the vesicle composition, size, and lipid packaging (Figure 2).

The α-syn shows a higher affinity for synaptic-like vesicles composed of negatively charged phospholipids, particularly phosphatidyl-glycerol and PS [104,105]. Of note, other components such as PC, PE, and PI, as well as cholesterol, sphingomyelin, and hexosylceramide are part of the SV membranes [106]. Moreover, an in vitro study performed on vesicles composed of anionic lipids 1-palmitoyl-2-oleoyl-sn-glycero-3-phosphoserine (POPS), 1-palmitoyl-2-oleoyl-*sn*-glycero-3-phosphoglycerol (POPG), or 1-palmitoyl-2-oleoyl-*sn*-glycero-3-phosphate (POPA) in 1:1 mixed with the zwitterionic 1-palmitoyl-2-oleoyl-*sn*-glycero-3-phosphocholine (POPC) shows that α-syn preferentially binds POPA with a 60-times higher affinity than POPS and POPG and very low affinity for POPC, confirming the importance of negatively charged lipid in α-syn binding (Figure 2b) [107].

α-syn binds preferentially to SUV rather than large unilamellar vesicles (LUV) of the same composition, most likely due to differences in phospholipid packing on the vesicle surface [107]. α-syn also shows an intrinsic affinity for highly curved lipid surfaces, which can be modulated by specific lipid components and the presence of bilayer defect. Other properties of the lipid bilayer could affect the α-syn binding including charge and surface hydrophobicity [108]. The interaction of α-syn with SUV composed of DOPE, DOPS, and DOPC favours the conformation of α-syn with the N-terminal region attached to the SUV and the region 65–97 available to bind another vesicles (Figure 2c) [62].

In addition, α-syn post-translational modifications could deeply affect the lipid interactions. For instance, α-syn acetylation increases the lipid-binding affinity [109] and specifically the acetylation of N-terminal α-syn is able to enhance binding to PC micelles and SUV with high curvature (16–20 nm) [110]. Phosphorylation of residue S129 increased or reduced the lipid-binding affinity of A30P and A53T, respectively [111]. Moreover, α-syn phosphorylation at residue Tyr39 could affect the α-syn conformation and, thus, the ability to bind lipids [112].

3. α-Synuclein Function in Exocytosis

The presynaptic localisation and the association of α-syn with lipids and the co-localisation of α-syn with proteins involved in exocytosis, such as Rab protein family members and soluble N-ethylmaleimide-sensitive factor attachment protein receptors (SNAREs), support the involvement of α-syn in synaptic plasticity and synaptic vesicle regulation [35]. Trafficking of SV is a process characterised by different steps including formation of the vesicles, tethering, docking, and fusion [113]. SV cluster at the presynaptic membrane and are then released by exocytosis, enabling communication between neurons.

It has been demonstrated that α-syn plays an active role in different processes occurring at the membrane during membrane fusion, membrane curvature during vesicle formation, docking, pore formation, regulation of neurotransmitter release, and vesicle recycling (Figures 3a,b and 4).

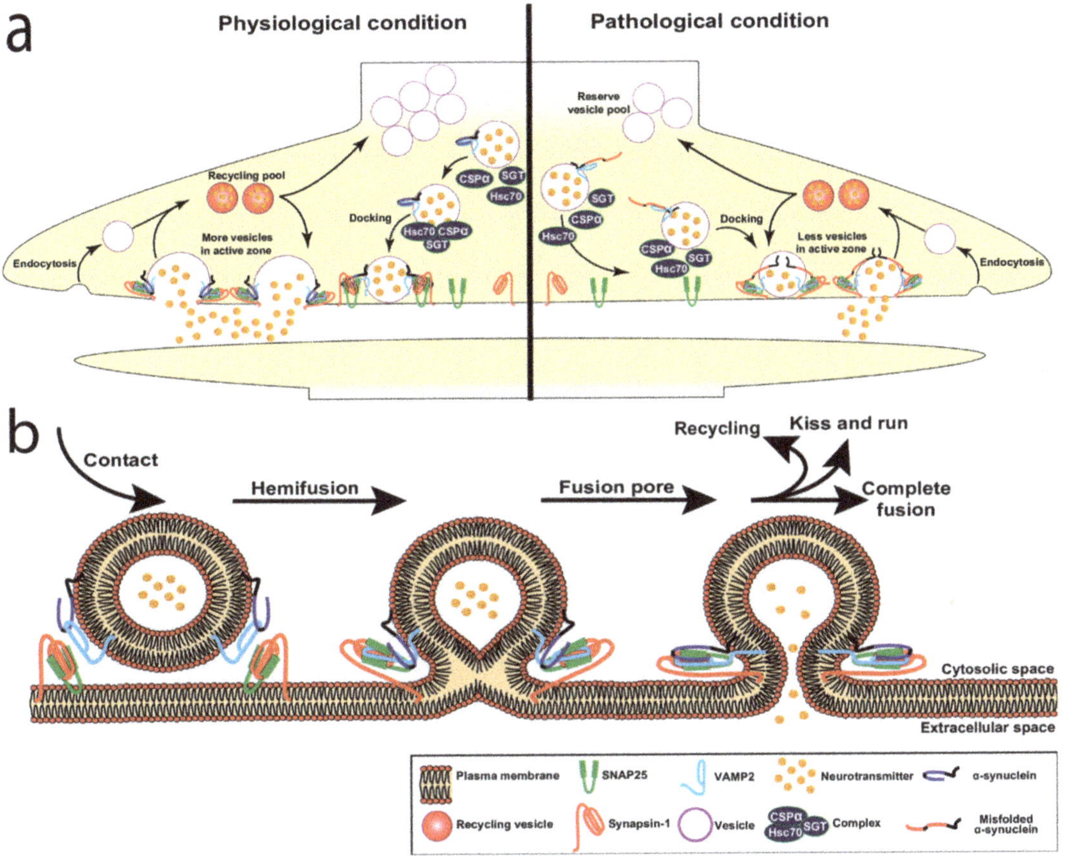

Figure 3. Schematic hypothesis of the role of α-synuclein (α-syn) in exocytosis. (**a**) α-syn, under physiological condition (**left** panel), interacts with the soluble N-ethylmaleimide-sensitive-factor attachment protein receptor (SNARE) vesicle-associated membrane protein 2 (VAMP2) on the synaptic vesicles (SV) surface, drives the docking of the SV to the active zone and regulates the formation of the tripartite SNARE-complex. Others synaptic partners including synapsin-1 and complexin act in the complex stabilisation. The SNARE-complex regulates the fusion of the SV with the synaptic membrane. After cargo release, the vesicles are recycled. Under pathological condition (**right** panel) aberrant forms of α-syn have a stronger binding affinity for VAMP2. The reduced availability of unbound VAMP2 molecules inhibits the SNARE complex formation and reduces the number of vesicles in the active zone. (**b**) α-syn actively participates in exocytosis by regulating SNARE complex formation and vesicle fusion events. Indeed, α-syn favours dilatation/closure of the fusion pore as well as regulates the kiss and run exocytosis. SNAP25 = synaptosome associated protein 25, CSPα = cysteine-string protein-α, Hsc70 = heat shock cognate 70, SGT = small glutamine-rich tetratricopeptide repeat-containing protein α.

3.1. α-Synuclein and Vesicle Docking

Presynaptic terminals contain hundreds to thousands of SV representing a reserve pool. Docking at the presynaptic PM is a crucial step that allows the physical contact of the vesicles with specialised areas of the presynaptic PM called active zones. When the vesicles initially dock, they are not competent for fusion. A vesicle priming step is therefore necessary to achieve a release-ready state upon calcium elevation and next fusion of the vesicles to the PM can take place. The docking is a highly regulated process that requires the interaction of two proteins located on the membrane of SV, vesicle-associated membrane protein 2 (VAMP2) and synaptotagmin and two PM proteins, syntaxin1 and synaptosomal-associated protein 25 (SNAP25) [114]. Although this protein complex is necessary for vesicle docking, α-syn and its interaction with lipids play an import role in this process (Figure 3a).

Interestingly, Man et al. quantified the stabilisation of synaptic-like vesicles docking to the PM by α-syn using total internal reflection fluorescence (TIRF) microscopy [38]. They discovered that, with a constant concentration of synaptic-like vesicles and varying concentrations of α-syn, the number of vesicles docking to the IPM surface increased with increasing levels of α-syn (a mean of 27 synaptic-like vesicles at 10μM of α-syn compared to 11,5 in the absence of α-syn) [38]. In addition, the estimated residence time for docking synaptic-like SUV doubled at 10 μM α-syn compared to the absence of α-syn. Because of the concentration effect, the authors suggest that several α-syn molecules may contribute to the stabilisation of the docking of a single vesicle. They also tested whether these changes would affect the mechanism of stabilisation of the synaptic-like SUV docked to the IPM surface and found that the synaptic-like vesicles docked to the IPM surface are strongly stabilised by α-syn probably also related to an increase in the amount of α-syn bound to IPM-GM than for IPM alone (see Section 2.2). These data indicate that modifications of the IPM composition may affect the mechanism of stabilisation of the docked vesicles by α-syn. In addition, cholesterol, which accounts for 31% of total lipid components of synaptic vesicles membranes [106], is an important regulator of α-syn membrane binding affinity. Indeed, the presence of cholesterol in the lipid bilayer reduces the affinity of the α-syn region 65–97 for synaptic-like vesicles. The in vitro study shows that, as a result, the overall affinity of α-syn for membrane is reduced and exposure of the unbound α-syn region 65–97 to the solvent leads to an increase in vesicle–vesicle interaction promoted by α-syn. Thus, cholesterol has a significant effect in vesicles clustering in vitro (Figure 2c) [62].

3.2. α-Synuclein and Fusion Pore

The fusion pore is one of the intermediate states during the fusion reaction when the vesicle connects to the PM which allows the release of the vesicle contents to the external medium. The fusion pore has a pronounced membrane curvature and is a highly dynamic structure (Figure 3). After opening, it reverts to the closed state or dilates leading to the full fusion with the PM, so that it can open and close several times before releasing or dilating further. A pore that closes after transient fusion leads to recapture of almost intact vesicles In contrast, regeneration of the vesicles is needed when vesicles fully fuse with the PM in order to maintain the vesicle pool. The vesicle recycling rate is thus an important event in maintaining the homeostasis of exo- and endocytosis mechanisms. In addition, the size of the pore is also an important parameter that controls the release depending on the nature of vesicle cargo as well as the strength of the stimulation. For instance, neuropeptides contained in large dense-core vesicles require a strong stimulation to be released [115], but small SV regulate the release of neurotransmitter via rapid flickering of the fusion pore [116].

The concept of pore formation for amyloid proteins was described earlier in 1993 in the Alzheimer's disease field by the description of annular shaped oligomers formed by the amyloid Aβ proteins and tau [117], which profoundly influence cellular homeostasis. The existence of an amyloid pore exerting its toxicity through the formation of ion channel pores disrupting the intracellular Ca^{2+} homeostasis was confirmed for Aβ in living cells most recently in 2017. Bode et al. demonstrated that Aβ oligomers, but not monomers and fibres, form ion channels that are toxic in cells [118]. The proportion of pores accounted only for one-third of the oligomer preparation. Thus, the authors suggested that among the potential mechanisms leading to the preferential channel formation, the importance of lipid composition specifically GM and cholesterols for Aβ insertion into the membrane could be an explanation [119–122]. This concept of lipid composition of membranes influencing the insertion of protein oligomers into membranes emerged also concerning the role of α-syn in PD, with the discovery by the Lansbury's group of membrane permeabilisation by a pore-like structure formed by annular shaped oligomers [123]. Indeed, α-syn oligomers penetrate in the membrane bilayer and give rise to an annular oligomeric species similar to a pore that acts as a protein channel. This formation of a ring-like structure has been confirmed using different sizes of α-syn oligomers, and this process has been directly associated with an increase in neuronal permeability [124].

More recently, in vitro studies demonstrate that α-syn participates in the fusion pore formation (Figure 3b) by penetrating into membranes and giving rise to the formation of annular pore-like structures that increase cell permeability and calcium influx [125]. The authors observed that α-syn affected the fusion pore. Upon α-syn overexpression, an accelerated release is observed preventing the pore closure. Conversely, the loss of α-syn has an opposing effect. Thus, there is a direct relationship between the level of expressed α-syn and the pore dilatation [126]. The ability to expand the fusion pore is not specific to α-syn, since the other synuclein isoforms, the β- and γ-synucleins, share this feature. Overall, this study shows that α-syn facilitates the exocytosis of secretory vesicles by increasing the rate of dilation of the fusion pore and the subsequent collapse of the vesicle membrane upon fusion at the active zone of the synapse [125]. This study is also in line with others showing that overexpression increases the rate of peptide discharge [127]; that α-syn has a similar effect on exocytosis of large dense core vesicles in neuronal cells or in PC12 or chromaffin cells [128]. Interestingly, while it was first suggested that α-syn mutations display little effect on exocytosis [128,129], the authors found a selective inhibition of the fusion pore by the mutations A30P and A53T linked to PD, as both mutations failed to accelerate peptide release in these experiments.

Although the role of α-syn on dilation of fusion pore has been established, some studies show that the formation and expansion of fusion pore are dynamic processes involving changes in membrane curvature, itself regulated by the SNARE protein complex.

3.3. α-Synuclein and the Cooperation with SNARE Proteins in Exocytosis

Fusion and exocytosis events require the regulated cooperation of α-syn with other synaptic proteins. In order to achieve the membrane fusion, membranes must overcome energy barriers created by charge repulsion and local dehydration of polar phospholipid head groups and by membrane deformation. The main actors mediating these processes are the SNAREs, main constituents of the SNARE complex to release energy, thereby enabling the bridging of the two membranes in close proximity. This phenomenon leads to the catalysis of membrane fusion (Figure 3a,b).

Several proteins contribute to regulate the SNARE complex, including α-syn. The SNARE complex mediating the fusion of SV with presynaptic PM during neurotransmission is composed of the target-SNAREs (t-SNAREs) Syntaxin-1 and SNAP25, located on the PM and the vesicular-SNARE (v-SNARE) synaptobrevin2/VAMP2 located in the membrane vesicles [130]. α-syn plays a crucial role in stabilising this complex. Burré et al. show that α-syn directly binds to the VAMP2 N-terminal domain through a short sequence in its C-terminal domain (residues 96–100). In support of this, it has been shown that α-syn lacking the VAMP2 protein–binding region (residues 1–95) does not interact with VAMP2 [131]. Similarly, bimolecular fluorescence complementation assays on hippocampal neurons confirm that the α-syn-VAMP2 interaction occurs at the synapse [131]. Moreover, the simultaneous interaction of monomeric α-syn with the acidic membrane lipids induces stabilisation of the tripartite SNARE complex [132]. These studies confirm the crucial role of α-syn in the stabilising the synaptic SNARE complex Syntaxin-1, SNAP25 and VAMP2 at the fusion pore. This evidence supports the role of α-syn as a chaperone of SNARE proteins. This notion is also supported by experiments performed on aggregated forms of non-mutated α-syn, which exhibit an enhanced VAMP2 binding affinity. The consequent increase of the fraction of VAMP2 bound to α-syn and the reduced amount of free VAMP2, reduce the formation of the SNARE complex inhibiting the docking of vesicles to the presynaptic terminal and impairing neurotransmission [133].

Beside the SNAREs other important protein partners are involved in the regulation of fusion event and interact with α-syn (Table 3). Another key regulator of SV trafficking is the Cysteine-String Protein-α (CSPα also known as DNAJC5). The DNAJ domain of the CSPα protein carries out its function by regulating the ATPase activity of the Heat Shock Cognate 70 (Hsc70). CSPα is a presynaptic protein that contributes to the stabilisation of the tripartite SNARE complex in a different way to α-syn. CSPα in complex with Hsc70 and the adaptor protein small glutamine-rich tetratricopeptide repeat-containing protein α (SGTA) acts as SNARE-chaperone, maintaining SNAP25 in the conformational state allowing the formation of SNARE complex [134]. It is interesting to note that genetic variants of the DNAJC family including CSPα/DNAJC5 have been associated with parkinsonism highlighting a functional pathway involved in the disease [135]. Another family of proteins acting at the synapse called synapsins interacts with α-syn and promotes α-syn functions at the synapse [136,137]. Synapsin III plays an important role as a cytosolic regulator of SV mobilisation [136]. In particular, synapsin regulates vesicle motility by influencing the targeting of α-syn to SV. Furthermore, complexin is another synaptic protein involved in the regulation of SNAREs in vivo and in neurotransmitter release through its interaction with SV [138]. The complexin is normally associated with the curved membrane [139] with a high packing defect [140].

Table 3. Synaptic proteins and their relationship to α-synuclein (α-syn) aggregates and vesicular alterations. The first column mentions the synaptic proteins while the second column mentions the models in which these proteins were studied. The third and fourth columns describe positive or negative effects observed on α-syn and vesicular functions, respectively.

Protein	Model	Positive Effect	Negative Effect
Complexin	Mice model over-expressing α-syn		Reduction in complexin 2 level in brain extracts from α-syn transgenic mice compared to controls [129].
	Mice α/β-syn double-KO		30% increase in complexins in α/β-syn double-KO mice [141].
CSPα	CSPα-KO mice		Reduction in SNAP25, Hsc70 and Hsp70 [141]. Impairment in SNARE complex formation [141].
			Increase in SNAP25 ubiquitylation and proteasomal degradation [134]. Reduction in SNAP25 [134,142] and Hsc70 protein levels [134] Impairment in SNARE-complex assembly [142].
	Neurons overexpressing CSPα.	CSPα suppresses the degradation of SNAP25 and Hsc70 and increases their protein levels [134].	
	CSPα-KO mice overexpressing WT or A30P α-syn.	Overexpression of WT α-syn but no A30P rescues the SNARE-complex assembly deficit induced by CSPα-KO [142].	
	Mice expressing a truncated human α-syn (1–120) injected with viral CSPα.	Viral CSPα injection reduces α-syn aggregates [142].	
SNAP25	*Snap25*$^{S187A/S187A}$ KI mice carrying an unphosphorylated form of SNAP25 (Ser/Ala phospho-dead mutation on position 187).		Increased number of endogenous α-syn aggregates associated with cytoplasmic side of the plasma membrane. Decreased ability of the SNARE complex assembly [143].

Table 3. *Cont.*

Protein	Model	Positive Effect	Negative Effect
Synapsin III	AAV-human α-syn injections in synapsin III KO mice.	Reduction in α-syn aggregation [144]. Reduction in the α-syn S129 phosphorylation in synapsin III KO mice in the striatum ipsilateral of an unilateral injection of AAV-human α-syn and no difference in the contralateral striatum [144].	
	Primary rodent dopaminergic neurons synapsin III KO.	Prevention of α-syn aggregation [135].	
	LB-enriched protein extracts from the SN of PD versus control brain samples.		LB-enriched fractions are immunopositive for both synapsin III and α-syn aggregates [145].
VAMP2	Rat cortical neurons treated with α-syn aggregates.		Direct binding of VAMP2 with α-syn aggregates. Reduction in VAMP2 and SNAP25 protein level, but no change in Syntaxin1A. 45% decrease in glutamate release [133].

Legend. α-syn = α-synuclein, α/β-syn = α/β-synucleins AAV = adeno-associated viral vector, CSPα = cysteine-string protein-α, Hsp70 = heat shock protein 70, Hsc70 = heat shock cognate 70kDa protein, KI = knock-in, KO = knockout, LB = Lewy body, SNAP25 = synaptosomal-associated protein 25, SNARE = soluble N-ethylmaleimide-sensitive-factor attachment protein receptor WT = wild type, VAMP2 = vesicles associated membrane protein.

Overall, these data demonstrate that α-syn requires the interaction with both lipids and numerous protein partners in order to fulfil its physiological synaptic function (summarised in Table 3). These interactions affect the localisation of α-syn at the synapse and its ability to stabilise the SNARE complex. In different models, the above-mentioned synaptic proteins involved in the SNARE complex formation and regulation could affect the aggregation state of α-syn as well as synaptic events in different manners as reported in Table 3, demonstrating the fundamental role of neuronal α-syn regulation in the pathogenesis of PD. Thus, pathogenic forms of α-syn altering these key interactions may result in altered SV trafficking and neurotransmitter release.

3.4. Loss and Overexpression of α-Synuclein in Neurotransmitter Release

Several experiments silencing or overexpressing α-syn levels have been conducted to demonstrate that α-syn acts as a modulator of release of several different neurotransmitters. Mice with α-syn KO show impaired regulation of the synaptic resting pool, but not the readily releasable pool [146]. As postulated by Senior et al. α-syn may be a negative regulator of neurotransmitter release, controlling both the rate of transfer of vesicles to the readily releasable pool and the probability of vesicle fusion to a given synaptic stimulation. In this study, the loss of α-syn in KO mice is suggested to cause an increase in probability of dopamine release from dopaminergic synapses [147]. Triple KO mice deficient in the proteins of the synuclein family (α-, β-, and γ-synucleins) show that synucleins are important factors to determine the synapse size [148]. Guo et al. also demonstrate that α-syn regulates the dopamine transporter named vesicular monoamine transporter 2 (VMAT2). SNCA KO models increase the concentration of VMAT2 molecules per vesicle [149], while overexpression inhibits the VMAT2 activity leading to increased cytosolic dopamine levels [150]. The activity of other neurotransmitter regulators such as the dopamine transporter (DAT) is also affected by α-syn. [151]. Indeed, WT α-syn interacts through the NAC domain (residues 58–107; Figure 1) with a region in the C-terminal (residues 598–620) of DAT [152,153]. Overexpression of α-syn has been suggested to induce an increased trafficking of DAT from the plasma membrane surface to the cytosol, where it has become toxic due to its ability to induce oxidative damage. In contrast, the overexpression of α-syn leads to a decrease of vesicle density and a reduced dopamine release. Such defects would in turn promote motor deficits [154,155]. Two potential mechanisms could explain such results: α-syn overexpression may (i) affects either the exocytosis or endocytosis of the recycling pool, or (ii) decreases the availability of the vesicle pool. Interestingly, the physiological role of α-syn in dopamine release has recently been better understood based on data obtained in mouse models by Somayaji et al. [156]. They demonstrated that α-syn promotes the dopamine release when neurons in the *substantia nigra* undergo action potential bursts separated by short intervals, in the range of few seconds. The authors suggest that the rapid facilitation may be associated with increased docking and fusion of SV to the membrane of active zones during exocytosis. Conversely they also demonstrated that a longer interval between two consecutive induced bursts in the range of minutes, is responsible for a depression of dopamine release that is α-syn-dependent. They proposed that this depression is due to synaptic exhaustion (Figure 4) This α-syn induced presynaptic plasticity is independent on calcium, but depends on the type of neuronal activity [156]. Thus, the authors propose that the dopamine release is strongly dependent on pore size and dilatation as well as on the α-syn protein expression level. In contrast, the release of other neurotransmitters, such as glutamate, is not affected by altered α-syn expression [157]. Additional information on deletion or overexpressing patterns of *SNCA* models are presented in Table 4.

Table 4. Fine deregulation of α-synuclein (α-syn) in exocytosis. In the table are presented divergent studies describing the role of α-syn as inhibitor or promoter of SV exocytosis. The studies are classified according to the models used and include in cellulo or in vivo α-syn overexpression and α-syn knockout models, as well as in vitro models using recombinant α-syn or artificial membrane vesicles and assays.

α-Synuclein Models	Positive Effect on Exocytosis	Negative Effect on Exocytosis
α-syn overexpression models		
PC12 cells and chromaffin cells overexpressing α-syn.		Reduced catecholamine release in both PC12 and chromaffin cells. Accumulation of docked vesicles at the plasma membrane in PC12, but not in chromaffin cells Potential inhibition of the priming of neurosecretory vesicles in chromaffin cells [128].
Transgenic expression of α-syn in CSPα knockout mice.	Rescue the assembly and function of the exocytic SNARE29, preventing neurodegeneration [159].	
Hippocampal neurons overexpressing α-syn.	Enhanced both spontaneous and evoked neurotransmitters release [160].	
Primary rat hippocampal neurons overexpressing α-syn and endogenous α-syn.	Promoted dilation of the fusion pore [125].	
α-syn KO/deletion models		
α-syn KO mice obtained by deleting exons 1 and 2 of the *SNCA* gene.	No impairment in structure of synapse, release of neurotransmitters, mobilisation of SV [141].	

Table 4. *Cont.*

α-Synuclein Models	Positive Effect on Exocytosis	Negative Effect on Exocytosis
α-syn KD by antisense oligonucleotides in hippocampal neurons.		Blocking the potentiation of synaptic transmission
α-syn KO mice obtained by deleting exons 4 and 5 of the *SNCA* gene in embryonal stem cells.		Dramatic loss of reserve vesicles and an increase in synaptic depression [146].
Non-viral gene therapy based on a new indatraline-conjugated antisense oligonucleotide (IND-ASO) to disrupt the α-synuclein mRNA transcription selectively in monoamine neurons of a PD-like mouse model and elderly non-human primates.	Intracerebroventricular and intranasal IND-ASO administration for four weeks in a mouse model with AAV-mediated WT human α-syn overexpression in dopamine neurons prevented the synthesis and accumulation of α-syn in the connected brain regions, improving dopamine neurotransmission [161].	
α-syn aggregates models and recombinant α-syn treatment		
Introduction of α-syn aggregates into single dopaminergic neurons via the patch electrode.		Accumulation of α-syn aggregates may chronically activate KATP channels leading to loss of excitability and dopamine release [162].
Synapse treated with recombinant human α-syn-112.		α-syn-112 strongly inhibits SV recycling [163].
Giant Lamprey synapse injected with α-syn.	Accumulation of clathrin coated pits and clathrin coated vesicles [164].	
In vitro studies		
Immobilised α-syn on sepharose beads incubated with radioactive arachidonic acid.		α-syn inhibits both exocytosis and SNARE complex formation by decreasing the levels of free arachidonic acid available to the SNARE proteins [165].
Single-vesicle and bulk in vitro lipid-mixing assays with α-syn purified monomer.	The α-syn monomers promote SNARE complex formation [166].	
Single-vesicle and bulk in vitro lipid-mixing assays with α-syn purified oligomers.		Interaction of large α-syn oligomers with VAMP2 Inhibition of SNARE complex formation Inhibition of docking vesicles [166].
In vitro lipid-mixing assay with monomers and oligomers.		Both α-syn monomers and α-syn oligomers induce the clustering of SV. The α-syn mutant T44P/A89P with reduced lipid-binding affinity reduces the clustering of SV by α-syn oligomers in vitro [167].

Legend. α-syn = α-synuclein, AAV = adeno-associated virus, CSPα = cysteine-string protein-α, KATP channel = ATP-sensitive potassium channel, KD = knockdown, KO = knockout, SNARE = soluble N-ethylmaleimide-sensitive factor (NSF) attachment protein receptor, SV = synaptic vesicles, VAMP2 = vesicle associated membrane protein 2, WT = wild type.

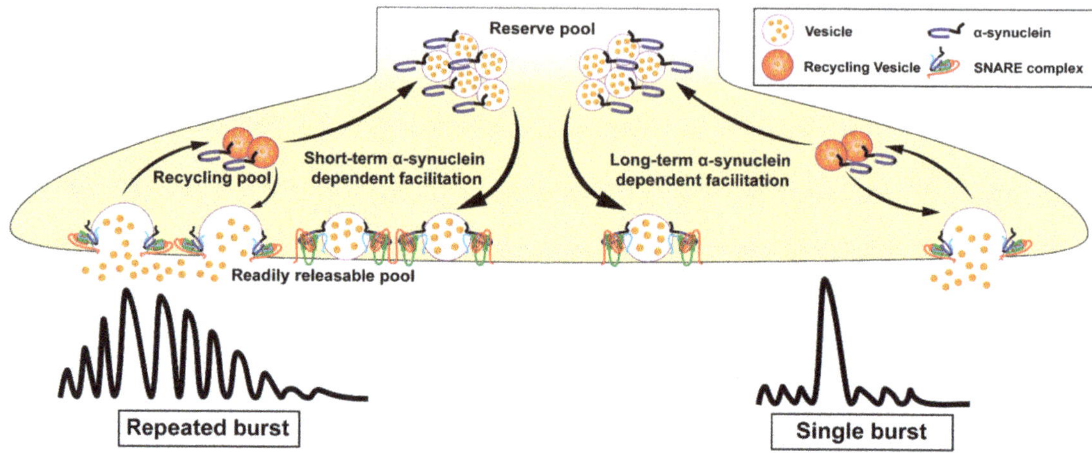

Figure 4. Exocytosis events mediated by α-synuclein (α-syn) are influenced by action potential bursts. α-syn influences exocytosis in different ways depending on the duration of action potential bursts. It is known that high frequency stimulation is responsible for the exhaustion of dopamine storage pool [158]. Dopamine release is promoted by α-syn when action potential bursts are separated by short intervals or reduces release when the interval between consecutive bursts is in the range of minutes [156].

3.5. Vesicle Recycling

The neurotransmitter release is a rapid and constant process that continuously requires the availability of newly formed SV. Although de novo synthesis of new SV occurs in the cell body, the main process that ensures the availability of SV pool is the vesicles recycling, a process in which the SV, after the exocytosis and the release of their cargo in the extracellular space are recycled by the cells through the fusion with the PM and the endocytosis. As mentioned previously, α-syn overexpression inhibits exocytosis, but the recycling of SV is also altered [168]. Indeed, this negative effect is mainly associated with the ability of dimers of α-syn to cluster SV leading to reduced vesicles mobilisation which blocks vesicle recycling at the PM [169].

Recently, an in vivo study also showed that the α-syn-112 isoform, produced by in-frame excision of exon 5, inhibits SV recycling. This inhibition is associated with the increased affinity of α-syn 112 for phospholipid binding and enhanced tendency to oligomerise. The same inhibitory effect has been found for α-syn-140 and the α-syn mutant A53T, particularly upon increased synaptic stimulation resulting in loss of SV and expansion of the PM [164].

3.6. Aberrant α-Synuclein in PD-Lipid Binding and Synaptic Function

The maintenance of a physiological and ordered α-syn conformation is among the parameters that influence its lipid-binding properties and functions, as mentioned above. Indeed, it is proposed that the pathological oligomerisation of α-syn and the formation of α-syn protofibrils lead to synaptic dysfunctions and neurotoxicity [170,171]. The conformation and folding of α-syn influence the behaviour and function of α-syn at the synapse. Although mutations of α-syn can affect its folding, lipid binding, and consequentially its function, there is evidence to suggest that α-syn dysfunctions at the synapse may be an early step in pathogenesis of PD [172–175], but the exact mechanism leading to pathology remains still unknown.

3.6.1. Oligomerisation of Pathogenic α-Synuclein and Lipid Binding

The point mutations associated with PD could promote oligomerisation and/or aggregation of α-syn by inducing alterations in the secondary structure and, thus, affecting lipid binding properties [176]. PD-related missense mutations are mostly located in the N-terminal region that interacts directly with lipid membranes. In vitro studies show that among the different pathogenic *SNCA* mutations (Figure 1), the A30P is a mutant defective in binding to phospholipids in membrane vesicles, while the A53T mutation has no effect on lipid binding [45,177]. Although the majority of α-syn mutations occurs in this membrane-binding site, not all have a reduced affinity for membrane binding, thus the effect of the A30P mutant is probably due to the presence of the proline residue, which is an amino-acid known to favour destabilisation of the α-helix secondary structure formation [45]. In vivo experiments corroborate these data. The A30P mutant reduced the α-syn interaction with membranes in rat isolated vesicles [178]. Furthermore, high frequency stimulation is responsible for depleting the dopamine storage pool. Interestingly, in mice overexpressing human α-syn A30P, a lower decline in dopamine release was observed after repeated stimulations compared to WT control mice. This effect is directly associated with the decrease in dopamine storage pool in A30P α-syn due to the faster exhaustion of dopamine storage pool compared to WT mice (Figure 4) [158]. These effects of A30P mutation could be explained by an alteration in the folding of α-syn protein leading to a closer association of the N- and C-termini in the mutant protein [179].

The A30P missense mutation as well as H50Q, G51D, A53E, A53T are also impacted by different intracellular environmental factors of which the physiological concentration of metals could affect the α-syn oligomerisation. In vitro, trivalent metal ions, such as $FeCl_3$ or $AlCl_3$, affect oligomerisation by increasing the A30P and decreasing the A53T and moderately decreasing α-syn H50Q, G51D, and A53E oligomer fractions compared to α-syn WT. No difference in oligomer formation was identified for the E46K mutant compared to the WT control [180]. In addition, an in vitro study using the membrane system dipalmitoyl-PC-SUV for which α-syn has strong affinity, shows that the lipid-binding of α-syn A30P and G51D is strongly and moderately reduced, respectively [180].

Since overexpression of α-syn through multiplication of its gene locus is a cause of PD, it is also interesting to note that α-syn overexpression through α-syn lentiviral injection induces a more severe phenotype and dopaminergic neuronal death. This overexpression contributes to increase the levels of some specific lipids such as oleic acid and unsaturated fatty acids [63,181]. In addition, the lipid composition favours or reverses the multimerisation of α-syn. In cell models stably expressing human WT α-syn or PD mutated α-syn, long chain polyunsaturated fatty acids (PUFA) promote α-syn multimerisation, while saturated fatty acids decrease α-syn multimers [79].

Thus, the α-syn mutations or α-syn multiplications tend to demonstrate the direct connection between lipids and α-syn oligomer seeding. Knowing that the cellular toxicity induced by α-syn oligomers correlates with their ability to disrupt synthetic and cellular membranes [170], this tends to support the notion of a pathological role of α-syn overexpression in PD. Indeed, Fusco et al. using two different types of α-syn oligomers show that the strain of α-syn oligomers, more prone to disrupt the lipid bilayer of synthetic membrane, localises in the luminal surface of artificial vesicles. In contrast, the α-syn oligomers not associated with cytotoxicity localise to the outer surface of the lipid bilayer [170].

However, the complexity of this relationship between oligomers and membranes is underlined by the recent observation that α-syn overexpression in yeast leads to lipid inclusions lacking the typical fibrillar form of α-syn that has since been considered as hallmark of synucleinopathies. Thereby, oligomerisation is not always observed in α-syn positive inclusion in PD brains. Immunostaining for α-syn in PD neurons shows the presence of irregularly shaped and diffuse inclusion structures, called pale bodies containing organelles and vesicles. Pale bodies have been considered as the first stage in the formation of a mature Lewy body [182]. In addition, the recent work on the composition of Lewy bodies brings out a new scenario supporting the hypothesis that PD is much more than a proteinopathy [41]. Indeed, Lewy bodies are mainly composed of damaged mitochondria, cytoskeletal components, phospholipids, sphingolipids, neutral lipids, lipid droplets (LD), cholesteryl esters, and α-syn oligomers [183]. Thus, this evidence leads to the hypothesis that membrane lipids may have a central role in the seeding, fibrillisation and accumulation of α-syn and that α-syn lipid cross-talk may be among the causes of Lewy pathology [181]. The reciprocal effect of α-syn and lipids points to the central role of both molecules in maintaining cellular homeostasis and probably synaptic functions.

In this context, the molecular cross-talk between lipids and α-syn needs to be further investigated in vivo in order to identify the key processes leading to synaptic dysfunctions.

3.6.2. Fine Regulation of α-Synuclein on Synaptic Activity

The lipid-dependent conformation and/or folding of α-syn influence(s) the α-syn behaviour and function at the synapse. Although the interaction of α-syn with the v-SNARE VAMP2 is well characterised, the exact role of α-syn in SNARE-dependent exocytosis at the synapse remains unclear since contrasting results show both positive and negative role of α-syn in SNARE regulation (Tables 3 and 4).

In favour of a positive role for α-syn in exocytosis, it has been shown that the conformational change from unfolded cytosolic monomer to the folded α-helical multimers renders α-syn capable of promoting the SNARE complex assembly by clustering VAMP2 molecules during SV docking [184]. Furthermore, the α/β/γ synuclein triple KO mouse model exhibits an impaired SNARE-complex assembly and a consequent loss of synaptic activity. This phenotype is reversed after overexpression of α-syn in α/β/γ synuclein KO neurons in an α-syn dose-dependent manner confirming the crucial role of α-syn in stabilising the SNARE complex [132]. Conversely, inhibitory effects of α-syn on exocytosis have also been described. Indeed, overexpression of α-syn inhibits neurotransmitter release by interfering with vesicle priming [128] or SV recycling [129]. Mice lacking α/β/γ synucleins show increased dopamine release associated with a reduced ability of the nerve terminals to store the vesicle pool. A reduced dopamine-content per vesicle was also detected, suggesting an important role of synucleins in dopamine regulation [185]. A study supporting inhibitory effects of α-syn on SNARE-complex assembly did not observe α-syn/VAMP2 interaction in purified synaptic terminals. Furthermore, the authors demonstrate that in vitro, α-syn reduces the level of arachidonic acid, an important regulator of the SNARE complex, thus affecting its formation and stabilisation [165].

Altogether, these divergent studies support the hypothesis that many factors and competitive interactions could regulate the state, the folding and the conformation of α-syn and thus its activity (Figures 3 and 5). Indeed, the differential affinity of α-syn regions (Figure 2) for different classes of lipids leads to the hypothesis that any metabolic dysfunction causing alterations in membrane composition, membrane GM content, or membrane lipid raft organisation could strongly affect the α-syn synaptic function and neurotransmission.

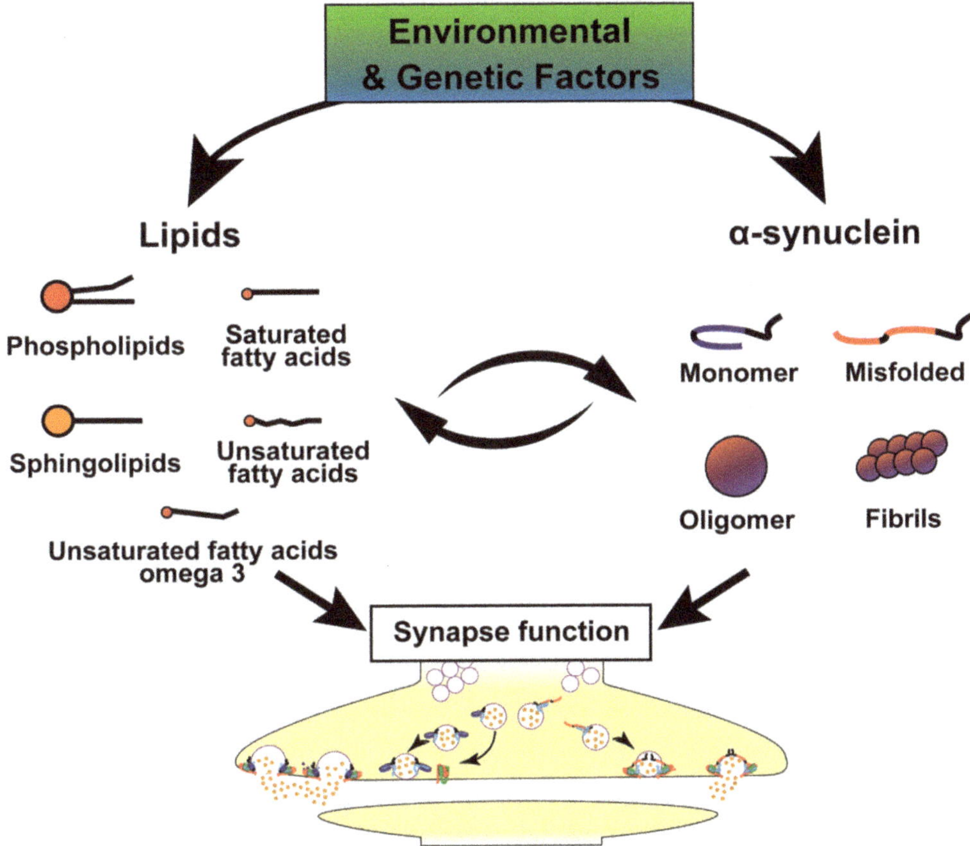

Figure 5. Schematic representation of genetic and environmental factors supporting the role of lipids in PD and synaptic homeostasis. The environmental factors include chemicals and toxins. Among the genetic determinants associated with α-synuclein (α-syn) pathology, many are associated with lipid metabolism or transport, such as *Ataxin2* gene (*ATXN2*), *Chromosome 19 Open Reading Frame 12* gene (*C19orf12*), *Galactosylceramidase* gene (*GALC*), *Glucosylceramidase* β gene (*GBA*), *Diacylglycerol Kinase Theta* gene (*DGKQ*), *ELOVL fatty acid elongase* gene (*ELOVL7*), *Phospholipase A2 group VI* gene (*PLA2G6*), *Scavenger Receptor Class B Member 2* gene (*SCARB2*), *Non-A4 Component Of Amyloid Precursor (SNCA)*, *Sterol Regulatory Element Binding Transcription Factor 1* gene (*SREBF-1*), and *Vacuolar Protein Sorting 13 homolog C* gene (*VPS13C*). Mutations in these genes are responsible for lipid alterations that can trigger the α-syn oligomerisation and consequentially compromise the α-syn synaptic dysfunctions. Aberrant forms of α-syn can also affect the lipids by modifying the membrane integrity [186]. Other parkinsonism-related genes link to vesicular trafficking includes *ATPase H+ Transporting Accessory Protein 2* gene (*ATP6AP2*), *ATPase Cation Transporting 13A2* gene (*ATP13A2*), *Parkinsonism Associated Deglycase* gene (*DJ1*), *DnaJ Heat Shock Protein Family (Hsp40) Member C6* gene (*DNAJC6*), *Leucine Rich Repeat Kinase 2* gene (*LRRK2*), *PTEN Induced Kinase 1* gene (*PINK1*), *Ras-Related Protein Rab-29* gene (*RAB29*), *Ras-Related Protein Rab-39B* gene (*RAB39B*), *Synaptojanin 1* gene (*SYNJ1*), *Synaptotagmin 11* gene (*SYT11*), and *VPS35 Retromer Complex Component gene (VPS35)* may be involved in deregulation of lipid homeostasis (see for review [157]). Physiological synaptic activity mediated by α-syn requires the co-operation of membranes and soluble interactors including lipidic components and protein partners (Table 3). Any disruption of the expression, localisation, interaction of α-syn and/or the above-mentioned partners can induce alterations at different levels of vesicle trafficking processes resulting in an altered neurotransmission and synaptic communication. Membrane phospholipids play an important role in this respect. Indeed, the α-syn-lipid interaction may represent an important step leading to conformational change and physiological multimerisation of α-syn [187]. It seems likely that any variation in membrane lipid composition or expression level of α-syn as well as the presence of α-syn mutant may compromise the α-syn binding properties and functions of α-syn.

4. Metabolic Alterations and Genetic Susceptibility Factors in PD, Implications for the α-Syn-Lipid Interplay

In light of the interplay between α-syn and lipids described in the previous section, it is interesting to verify what insight exists into this interplay in PD patients and models. As shown in Table 2, different classes of lipids are indeed deregulated in PD patient samples and PD animal models leading to pathological alterations of α-syn. Furthermore, the interaction of α-syn with lipids is important for α-syn to interact with synaptic protein partners. As a known example, PS has been shown to regulate the α-syn-mediated docking of SV by facilitating the formation of the SNARE complex. The PUFA are a class of lipids actively involved in SV trafficking and their interaction with the N-terminal segment of α-syn increases the α-syn oligomerisation [65]. Thus, alterations in membrane lipid components are widely observed in PD and, as described above, these data confirm their central role in the maintenance of cellular homeostasis.

Several enzymes involved in lipid metabolism also display abnormal activities in biofluids or brain tissues from PD patients or cellular models (Table 2). An increase in sphingomyelinase activity in PD brains has been reported and it has been associated with increased levels of ceramides that activate apoptotic processes. Inhibition of the enzyme sphingosine Kinase (Sphk1), involved in the regulation of sphingolipid homeostasis, correlates with enhanced secretion and propagation of α-syn. The phospholipase D1 enzyme (PLD1), involved in phospholipid hydrolysis is able to prevent α-syn accumulation by activating autophagic flux. Reduced activity and expression level of this enzyme are observed in post-mortem brain of *patients with Lewy body dementia* [87]. Alterations in glycosphingolipid metabolism are also identified in CSF and blood of PD patients as well as modulation of several lysosomal enzyme activities such as increased β-galactosidase and decreased β-hexosaminidase [91], contributing to the deregulation of lipid levels. In addition, some of the lipids deregulated in PD participate in pro-inflammatory processes (sphingolipids and long-chain ceramides) [188] or in anti-inflammatory phenotypes (short-chain ceramides) supporting the evidence that the above-mentioned metabolic alterations contribute to neuroinflammation, a known hallmark of PD [65,189]. Different mechanisms are involved including inflammasome activation, altered calcium homeostasis, changes in the blood–brain barrier permeability and recruitment of peripheral immune cells [91].

Moreover, several studies support a lipid dysfunction in PD that not only affects α-syn, but also actively participates in PD pathogenesis. This new hypothesis is supported by the recent advances in the genetic studies of PD/parkinsonism as well as susceptibility genes associated with α-syn deposition are involved in lipid metabolism as described in the Figure 5 and Table 5, thus shedding light on lipid alterations as important contributors or determinants of synucleinopathies. Moreover, several parkinsonism-related genes including *ATPase H+ Transporting Accessory Protein 2* gene (*ATP6AP2*), *ATPase Cation Transporting 13A2* gene (*ATP13A2*), *Parkinsonism Associated Deglycase* gene (*DJ1*), *DnaJ Heat Shock Protein Family (Hsp40) Member C6* gene (*DNAJC6*), *Leucine Rich Repeat Kinase 2* gene (*LRRK2*), *PTEN Induced Kinase 1* gene (*PINK1*), *Ras-Related Protein Rab-29* gene (*RAB29*), *Ras-Related Protein Rab-39B* gene (*RAB39B*), *Vacuolar Protein Sorting 13 homolog C* gene (*VPS13C*), *VPS35 Retromer Complex Component* gene (*VPS35*), *Synaptojanin 1* gene (*SYNJ1*), *Synaptotagmin 11* gene (*SYT11*) (see for review [157]) are actively involved in membrane and vesicle trafficking and are (or may indirectly be) associated with deregulation of lipid homeostasis supporting this view.

All of the above-mentioned metabolic and genetic dysfunctions contributing to development of PD or α-syn-related pathologies (Tables 2 and 5, Figure 5) emphasize the need to further investigate the interplay at the synapse between lipids and α-syn. This is all the more important as several studies tend to show that synaptic dysfunctions occur early in the development of disease [174,175]. Moreover, a reduction in dopamine release as well as alterations of proteins involved in the exocytosis of SV occur prior to the dopaminergic cell death [190]. Given that α-syn is a key determinant of several synucleinopathies, it is thus of great interest to further investigate these altered pathways in multiple system atrophy, dementia with Lewy bodies and Alzheimer's disease.

Table 5. Presentation of some genetic determinants associated with α-synuclein (α-syn) pathology and having a direct relationship with lipid pathways. This table provides some examples of different genes (first column) associated with α-syn pathology and/or to parkinsonism (second column). Mutations in these genes can directly affect the biological metabolism (third column) and, in some cases, the properties of α-syn (fourth column).

Genes	Genetic Determinants Associated with α-Syn Pathologies	Effect on Lipids	Effects on α-Syn
ATXN2	Diseases associated with *ATXN2* include Spinocerebellar Ataxia 2 and PD/parkinsonism with LB pathology [191].	Ataxin-2 expansion affects ceramide-sphingomyelin metabolism [192].	NI
C19orf12	*C19orf12* is associated with Neurodegeneration with Brain Iron Accumulation disorders with prominent widespread Lewy body pathology [193].	Role in lipid homeostasis [194].	NI
DGKQ	*DGKQ* emerged as PD risk factor in independent GWAS studies [195,196].	Controls the cellular content of diglycerides.	*DGKQ* loss-of-function in PD might potentially leads to enhanced transcription of *SNCA* [197].
ELOVL7	*ELOVL7* identified in GWAS studies as PD-associated gene [198].	FA elongase 7 plays a role in synthesis of long-chain saturated fatty acids involved as precursors of membrane lipids and lipid mediators [199].	Defects in very long chain fatty acid synthesis enhance the toxicity of α-syn WT, A53T and E46K toxicity in a yeast model of PD. The effect on α-syn A30P is inappreciable in a yeast model of PD [75].
GALC	Mutations in the *GALC* gene are responsible for Krabbe disease, a demyelinating disorder characterised by the presence of neuronal aggregates, in part composed of α-syn [200].	GALC catalyses the hydrolysis of substrates including galactosylceramide (GalC) and galactosylsphingosine. In PD patients, higher levels of galactosylsphingosine were found respect to controls [201].	Galactosylceramidase treatment improves the survival and health of KD mice, prevents the formation of α-syn in spinal neurons [200] galactosylsphingosine accelerates aggregation of α-syn in a dose-dependent manner [202].
GBA	PD risk factor confirmed in GWAS studies [203,204].	Involved in glycolipid catabolism.	The decreased GCase activity identified in CSF and blood PD patients and the consequent increase in glucosylceramide level directly correlates with increased α-syn oligomer formation.
PLA2G6	*PLA2G6* is causative for PARK14 in patients with autosomal recessive dystonia-parkinsonism [205].	PLA2G6 hydrolyses the sn-2 acyl chain of glycerophospholipids in free fatty acids and lysophospholipids [206].	Pla2g6 KO mouse neurons show early increase in α-syn/phospho α-syn level [207]. Increased expression of α-syn in cell and animal model with PLA2G6 dysfunction [207].
SCARB2	*SCARB2* locus identified in GWAS studies as PD-associated gene. This gene encodes LIMP2 [208].	LIMP2 deficiency can lead to a decrease in GCase activity and α-syn degradation [209].	LIMP2 deficiency can lead to a decrease in GCase activity and α-syn degradation [209].
SREBF1	*SREBF1* locus identified as PD risk factor in GWAS [208].	*SREBF1* encodes SREBP-1 that regulates synthesis of sterol.	NI

Table 5. *Cont.*

Genes	Genetic Determinants Associated with α-Syn Pathologies	Effect on Lipids	Effects on α-Syn
VPS13C	*VPS13C* first mutation identified in a form of early-onset parkinsonism; the pathological features were reminiscent of diffuse Lewy body disease [210].	VPS13C is a lipid transport proteins [211].	NI

Legend. α-syn = α-synuclein, ATXN2 = Ataxin2 gene, C19orf12 = Chromosome 19 Open Reading Frame 12 gene, CSF = Cerebrospinal fluid, DGKQ = Diacylglycerol Kinase Theta gene, ELOVL7 = ELOVL Fatty Acid Elongase 7, GALC = Galactosylceramidase gene, GBA = Glucosylceramidase β gene, GWAS = genome-wide association study, KD = knockdown, KO = knockout NI = no information, PLA2G6 = Phospholipase A2 Group VI gene, SCARB2 = Scavenger Receptor Class B Member 2 gene, SREBF1 = Sterol Regulatory Element Binding Transcription Factor 1 gene, VPS13C = Vacuolar Protein Sorting 13 homolog C gene.

5. Future Directions

5.1. Towards Further Fundamental Advances

Despite these exciting recent progresses, numerous questions remain to be resolved in order to better understand the interplay between α-syn and lipid membranes and their role at the synapse in the different steps leading to neurotransmission. Indeed, vesicle fusion events as well as transient interactions of intra-membranous proteins with cytosolic and cytoskeletal partners make the biological membrane a highly dynamic system, undergoing constant rearrangements during vesicle and membrane trafficking. Thus, artificial systems used in in vitro studies miss the complexity of biological membranes and do not take these parameters into account. Moreover, the physiological state of α-syn is influenced, as mentioned earlier, by various intra- and extra-cellular stimuli including temperature and pH variations, protein interactions, metal ion concentrations and ionic strength. Simplified artificial systems make it difficult to interpret all the combined parameters and hinder the extrapolation of results to in vivo or human models. Thus, future research should develop new tools capable of integrating the complexity of intracellular environment. The conformational change of α-syn, induced by interactions with membranes is transient and occurs rapidly during the physiological synaptic activity. Consequently, the membrane mimetic models should consider the dynamism of α-syn conformations. To date, most of the research has focused on the main α-syn 140 isoforms. However, several α-syn post-translational modifications (acetylation, phosphorylation, glycosylation, ubiquitin conjugation, etc.), as well as several types of α-syn isoforms exist and their relationships with different classes of lipids are still in their infancy [56]. The main α-syn isoforms in the brains are the α-syn 140 AA and the α-syn 112 AA, but there are others, including α-syn 98 or 66 AA, as well as the α-syn fragments including α-syn 1–96 and α-syn 65–140 identified in human brains [212]. Moreover, these α-syn fragments and truncated forms have been identified in other synucleinopathies including multiple system atrophy and dementia with Lewy bodies [213]. In addition, the effect of different types of oligomers in the interaction with membranes in both physiological and pathological conditions has yet to be deciphered. It will thus be interesting to define more precisely the lipid interplay with each of the different types of α-syn forms and post-translational modifications of α-syn.

In addition, altered lipid levels and metabolic pathways associated with PD and other synucleinopathies evolve with disease progression; in this respect, it is of interest to perform lipidomic analyses at different stages of the disease in order to assess the effect of lipid alterations on α-syn-dependent synaptic activity.

5.2. Towards Target Identification and Pharmacological Strategies

Different classes of lipids are actively involved in synaptic functions and, some of them affect the α-syn homeostasis by modulating its conformation, aggregation and finally its cytotoxicity (Table 2). In this regard, future strategies could emerge to modulate the impact of certain lipids in neurodegenerative disorders including synucleinopathies. To date, promising therapeutic approaches aim to modulate the levels of lipids by targeting the activities of proteins involved in the metabolism of the lipid pathway, including enzymes or lipid transporters. In addition, given the neuroprotective effects of some lipids, their direct administration is emerging as a promising strategy to alleviate α-syn cytotoxicity. Other therapeutic strategies point to alleviate conditions associated with PD caused by environmental factors. Finally, synaptic proteins have also been analysed as potential target for therapeutical strategies aiming to restore synaptic function in animal models of PD. Several examples for each of these approaches are detailed below.

5.2.1. Targeting Membrane Lipids or α-Synuclein Membrane Affinity

The α-syn-mediated toxicity in neuronal cellular models could be improved by inhibitors of the stearoyl-CoA desaturase enzyme [80]. The same effect is observed in cellular models overexpressing an engineered α-syn characterised by E35K + E46K + E61K mutations, which leads to the formation of round inclusions [214]. These observations suggest that inhibition of fatty acid desaturation could prevent the oligomerisation and α-syn-mediated toxicity. Based on such evidence, development of strategies to decrease the oligomerisation and aggregation of α-syn might be promising in particular for PD patients, heterozygotes or homozygotes for *SNCA* mutations, or gene multiplication. In this context, decreasing oleic acid production by stearoyl-CoA desaturase (SCD) inhibitors is emerging as potential strategy to rescue α-syn cytotoxicity [58]. An in vitro study also demonstrates a potential protective role of arachidonic acid, which is able to induce the formation of ordered and α-helical structured α-syn oligomers, resistant to fibrillisation [215].

Another option is to target lipids participating in synaptic functions, acting on GM which are identified as important factors in maintaining neuronal function [59] and because a consistent decrease in GM brain content has been observed in PD (Table 2). In addition, the intranasal infusion of GD3 and GM1 gangliosides alleviates α-syn toxicity and improves the function of midbrain dopaminergic neurons [216].

Interestingly, a recent approach aims to modulate the affinity of α-syn for membranes and, among them, a promising molecule has emerged known as the anti-microbial squalamine [217].

5.2.2. Environmental Factors and Potential Therapeutic Strategies

Interestingly, α-syn binds to LD, a lipid storage organelle contributing to energy metabolism [218,219]. Overexpression of α-syn in neuronal cells induces accumulation of LD which, in turn, increases the amount of proteinase K-resistant forms of α-syn, suggesting a potential pathological role of LD in PD [220]. Several environmental factors deeply affect lipid homeostasis and, among them, diet plays an important role. Indeed, dietary nutrients are the main substrates of the gut microbiota, which can process and metabolise them. Conversely, dietary nutrients can have an impact on the composition and metabolic activity of the gut microbiota [221]. These processes lead to the productions of intermediate metabolites that profoundly affect host energy homeostasis as well as glucose and lipid metabolism. Studies in animal models confirm that the cross-talk between the gut microbiota and the dietary lipids contributes to the regulation of lipid levels in biofluids and tissues [222].

In addition, increased LD formation in dopaminergic neurons [168] has been correlated with iron accumulation, a condition described both in PD patients and in animal models. Iron accumulation is also responsible for lipid peroxidation which, in turn, activates a caspase-independent cell-death pathway known as ferroptosis. Of major interest, pharmacological administration of the iron chelator deferiprone, reduces the abnormally high deposition of iron in the SN, as evidenced by a reduction in the progression of motor deficits in a clinical trial in early-stage PD [223]. The potential of this molecule as a PD modifier is currently being tested in an ongoing phase 3 clinical trial [224].

5.2.3. Targeting Synaptic Proteins

Although many therapeutic strategies aim to target lipids, other approaches directly target synaptic proteins. Among synaptic proteins, CSPα acts as anti-neurodegenerative molecule; this evidence is supported by studies in CSPα-deficient animal models showing impaired synaptic function. Of note, in CSPα-KO mice overexpression of α-syn WT rescues the SNARE-complex assembly deficit and this positive effect is associated with the ability of α-syn to interact with phospholipids. Indeed, the rescue of SNARE complex formation is not observed for A30P, the α-syn mutant with reduced phospholipid binding. These different observations tend to demonstrate the importance of CSPα in the prevention of neurodegeneration. The neuroprotective role of CSPα was reinforced by the demonstration from Spillantini's group [225] that viral injection of CSPα into transgenic mice expressing a truncated human α-syn (1–120) reduces α-syn aggregates. They first observed that α-syn aggregation at the synapse is associated with a decrease of CSPα, suggesting that α-syn synaptic aggregation affects the CSPα levels. Its function is also affected, as they observed a reduction in the CSPα/Hsc70 complexes with STGa in the striatum. In cellulo, they found that overexpression of CSPα rescues the alteration of vesicle recycling induced by α-syn overexpression. These data confirm that CSPα is an interesting synaptic target. Another target of interest is the synapsin III. The absence of this synaptic protein prevents the formation of α-syn aggregates in primary rodent dopaminergic neurons as well as a reduction in α-syn oligomers and a reduced level of α-syn S129 phosphorylation. Therefore, the loss of synapsin III displays protective effects on synaptic damage and neurodegeneration [144] and also confirms the central role of synapsin III on α-syn aggregation.

Altogether, these recent experiments provide optimistic perspectives in terms of potential targets for successful therapeutics for synucleinopathies.

6. Conclusions

All of the above-mentioned studies have clearly illustrated that, although the interactions between α-syn and lipids were identified from the first characterisation of the α-syn primary structure, the roles of lipids in the pathophysiology of PD have recently come to light in an insistent manner, like an elephant in the room. This central role of lipids is sustained by many pieces of evidence: (1) lipids and degenerated organelles represent the most abundant components of Lewy bodies [41] (Section 2); (2) α-syn shows a differential binding according to the compositions of inner and outer leaflets of plasma membranes (Section 2); (3) lipid membranes are directly involved in different steps of synaptic exocytosis (Section 3, Table 6); (4) an increasing number of genes associated with α-syn deposition is directly associated with lipid metabolism (Table 5, Section 4); (5) an increasing number of PD-genes affecting α-syn homeostasis is directly or indirectly related to lipid metabolism, such as those acting in vesicular and membrane trafficking (Table 5, Figure 5) [47]; (6) recent advances in therapeutic research show that lipid modulation can directly alleviate the α-syn pathology as well as the synaptic dysfunction (Table 2, Section 5). All these data illustrate the interplay between α-syn and lipids, and suggest that, at least under certain conditions, lipids could contribute to the development of the disease. For these reasons, it seems clear that lipids may contribute to the synaptic dysfunctions leading to PD, highlighting the need to better characterise the lipid/α-syn relationship in vivo. These exciting recent progresses, also point to numerous questions yet to be resolved in order to better understand the interplay among α-syn, lipid membranes, and their role at the synapse in the different steps leading to neurotransmission (Box 1).

Box 1. Unsolved issues going forward in the field of the interplay among α-syn, lipids, and their role at the synapse

What are the mechanisms by which pathological α-syn oligomers physically interact with lipid bilayer and affect cellular homeostasis? Different hypotheses have been proposed: (1) α-syn oligomers form an annular pore-like structure, similar to an ion channel, highly dynamic and capable of switching from open to close conformation and allowing the non-selective passage of ions with a consequent alteration of cellular homeostasis. (2) α-syn oligomers, when bound to membrane phospholipids by electrostatic interactions, cause the thinning of lipid bilayer with consequent membrane leakage. (3) Binding of α-syn oligomers to bilayer packing defects induces the extraction of phospholipids with consequent bilayer instability and degradation (4) α-syn trimers and tetramers induce the formation of a lipoprotein particles, called nanodiscs, which are ring shaped by their ability to wrap around the phospholipid bilayer [43].
Are the interactions between α-syn-oligomers and membranes identified in reconstituted systems translatable in vivo? Although the mimetic membrane systems have helped to define the affinity of single α-syn domain for specific classes of lipids (Figure 2), they do not consider the complexity of the biological system. There is therefore a need to further validate these interactions in in vivo models with more sensitive tools, and in the future, to integrate into these analyses several factors such as α-syn oligomers heterogeneity, size, intracellular amounts, kinetic transitions, post-translational modifications, and parameters membranes-related, such as phospholipid bilayer asymmetry and compositional change.
What is the effect of pathogenic α-syn oligomers on membrane homeostasis according to the stage of disease progression? Indeed, the α-syn oligomerisation, the metabolic alterations, as well as the variation in lipid content in brain and biological fluids, change over the time. Moreover, based on the central and direct role that phospholipids have in synaptic functions, it will be necessary to estimate the cytotoxic effect in vivo of α-syn oligomers on phospholipid bilayers. This will help to better elucidate the correlation between structural membrane alteration and PD pathophysiology and related disorders.

Table 6. Summary table on α-synuclein (α-syn) lipid interactions and synaptic dysfunctions in different models. The table provides the main information on α-syn wild type and PD-associated modifications (first column) and their effect on membrane lipids interaction (second column) that has been extensively described in Section 2. The mains functional effects of α-syn on the different steps of exocytosis are also described, highlighting the role of α-syn on the regulation of docking vesicles (third column, explained in the Section 3.1), fusion pore (fourth column, described in Section 3.2), exocytosis (fifth column, Sections 3.3 and 3.4), and vesicle recycling (sixth column, Section 3.5).

Type of α-Syn Modifications	Lipid Effect	Vesicle Trafficking			
		Docking	Fusion Pore	Exocytosis	Recycling
WT 140	Increased α-helical multimers formation.	Increased cluster of VAMP2- vesicles and SNARE complex assembly.			
A30P 140	Decreased membrane binding. Decreased membrane curvature. Abolition of interaction with lipid-rafts.	Accumulation of docked vesicles at the plasma membrane. Decreased priming of neurosecretory vesicles.	Perturbation of fusion pore formation.	Decreased catecholamine release. No change in synaptic exocytosis.	
A53T 140	Increased multimerisation long chain PUFA-mediated. Decreased multimerisation mediated by saturated fatty acids. No change in lipid binding.	Clustering of VAMP2 SV at the active zone.	Perturbation of fusion pore formation.		Perturbation of SV recycling.

Table 6. *Cont.*

Type of α-Syn Modifications	Lipid Effect	Vesicle Trafficking			
		Docking	Fusion Pore	Exocytosis	Recycling
E46K 140		Clustering of VAMP2 SV at the active zone.			
KO *		Decreased reserve pool.		Increased concentration of VMAT2 molecules per vesicle.	Increased dopamine release.
Overexpression	Decreased membrane curvature induction. Increased oleic acid and unsaturated fatty acid.	Decreased reserve vesicles. Decreased vesicle density.	Prevention the fusion pore closure.	Rescue the SNARE-complex assembly deficit in CSPα-KO mice. Decreased level of synapsins and complexins.	Increased cytosolic dopamine levels due to inhibition of VMAT2 activity.
Overexpression in CSPα-KO mice		Rescues the SNARE-complex assembly deficit.			Prevents neurodegeneration.
WT 112	Increased phospholipid binding Increased tendency to oligomerisation.				Perturbation SV recycling.
Soluble aggregates	Increased aggregates by PUFA.				
Pathological aggregates	Increased aggregation by cholesterol, lipids with short saturated acyl chain, GM1, and GM3.	Perturbation of vesicles docking at the presynaptic terminal.		Increased VAMP2 binding affinity.	Alteration in neurotransmission.

Legend. α-syn = α-synuclein, CSPα = cysteine-string protein-α, GM = gangliosides, KO = knockout, PUFA = polyunsaturated fatty acids, SNARE = soluble N-ethylmaleimide-sensitive factor (NSF) attachment protein receptor, SV = synaptic vesicles, VAMP2 = vesicle associated membrane protein 2, VMAT2 = vesicular monoamine transporter 2, WT = wild type. * see Table 4. The effect on exocytosis depends on the knockout model.

Author Contributions: A.S. and M.-C.C.-H. conducted most of the literature search, and drafted, reviewed, and edited the manuscript. J.-M.T. provided valuable discussions and reviewed the manuscript. A.M. drafted the figures. M.-C.C.-H. supervised the study. All authors have read and agreed to the published version of the manuscript.

Funding: The authors' research was financed by Inserm, CHU de Lille and Université de Lille, the ANR (Agence Nationale de Recherche, France) grant ANR-16-CE16-0012-02 MeTDePaDi, grant ANR-20-CE16-0008 Synapark, grant ANR-21-CE16 PARK-PEP, Fondation de France (Maladie de Parkinson, R19199EK), France Parkinson (R16008) and the Michael J. Fox Foundation, grant numbers 6709.03, 10255.03 and 12938.04.

Data Availability Statement: No new data were created or analysed in this study. Data sharing is not applicable to this article.

Acknowledgments: We wish to thank Cristine Alves Da Costa, the guest editor of the special review entitled "Hallmarks of Parkinson's disease", as well as Beatty Teng, for their support. We also wish to thank members of the LilNCog Research Centre, in particular the administrative team and the members of the BBC team, for their support of our research work.

Conflicts of Interest: The authors declare no conflict of interest. The funders had no role in the design of the study; in the collection, analyses, or interpretation of data; in the writing of the manuscript, or in the decision to publish the results.

References

1. Kieburtz, K.; Wunderle, K.B. Parkinson's disease: Evidence for environmental risk factors. *Mov. Disord.* **2013**, *28*, 8–13. [CrossRef]
2. Kalia, L.V.; Lang, A.E. Parkinson's disease. *Lancet* **2015**, *386*, 896–912. [CrossRef]
3. Braak, H.; Ghebremedhin, E.; Rüb, U.; Bratzke, H.; Del Tredici, K. Stages in the development of Parkinson's disease-related pathology. *Cell Tissue Res.* **2004**, *318*, 121–134. [CrossRef]
4. Croisier, E.; Moran, L.B.; Dexter, D.T.; Pearce, R.K.B.; Graeber, M.B. Microglial inflammation in the parkinsonian substantia nigra: Relationship to alpha-synuclein deposition. *J. Neuroinflamm.* **2005**, *2*, 14. [CrossRef]
5. Campion, D.; Martin, C.; Heilig, R.; Charbonnier, F.; Moreau, V.; Flaman, J.M.; Petit, J.L.; Hannequin, D.; Brice, A.; Frebourg†, T. The NACP/synuclein gene: Chromosomal assignment and screening for alterations in Alzheimer disease. *Genomics* **1995**, *26*, 254–257. [CrossRef]
6. Uéda, K.; Fukushima, H.; Masliah, E.; Xia, Y.; Iwai, A.; Yoshimoto, M.; Otero, D.A.; Kondo, J.; Ihara, Y.; Saitoh, T. Molecular cloning of cDNA encoding an unrecognized component of amyloid in Alzheimer disease. *Proc. Natl. Acad. Sci. USA* **1993**, *90*, 11282–11286. [CrossRef] [PubMed]
7. Spillantini, M.G.; Divane, A.; Goedert, M. Assignment of Human α-Synuclein (SNCA) and β-Synuclein (SNCB) Genes to Chromosomes 4q21 and 5q35. *Genomics* **1995**, *27*, 379–381. [CrossRef] [PubMed]
8. Polymeropoulos, M.H.; Lavedan, C.; Leroy, E.; Ide, S.E.; Dehejia, A.; Dutra, A.; Pike, B.; Root, H.; Rubenstein, J.; Boyer, R.; et al. Mutation in the -Synuclein Gene Identified in Families with Parkinson's Disease. *Science* **1997**, *276*, 2045–2047. [CrossRef]
9. Hoffman-Zacharska, D.; Koziorowski, D.; Ross, O.A.; Milewski, M.; Poznanski, J.A.; Jurek, M.; Wszolek, Z.K.; Soto-Ortolaza, A.; Awek, J.A.S.; Janik, P.; et al. Novel A18T and pA29S substitutions in α-synuclein may be associated with sporadic Parkinson's disease. *Parkinsonism Relat. Disord.* **2013**, *19*, 1057–1060. [CrossRef]
10. Liu, H.; Koros, C.; Strohäker, T.; Schulte, C.; Bozi, M.; Varvaresos, S.; Ibáñez de Opakua, A.; Simitsi, A.M.; Bougea, A.; Voumvourakis, K.; et al. A Novel SNCA A30G Mutation Causes Familial Parkinson's Disease. *Mov. Disord.* **2021**, *36*, 1624–1633. [CrossRef]
11. Brás, J.; Gibbons, E.; Guerreiro, R. Genetics of synucleins in neurodegenerative diseases. *Acta Neuropathol.* **2021**, *141*, 471–490. [CrossRef]
12. Singleton, A.B.; Farrer, M.; Johnson, J.; Singleton, A.; Hague, S.; Kachergus, J.; Hulihan, M.; Peuralinna, T.; Dutra, A.; Nussbaum, R.; et al. alpha-Synuclein locus triplication causes Parkinson's disease. *Science* **2003**, *302*, 841. [CrossRef]
13. Chartier-Harlin, M.-C.; Kachergus, J.; Roumier, C.; Mouroux, V.; Douay, X.; Lincoln, S.; Levecque, C.; Larvor, L.; Andrieux, J.; Hulihan, M.; et al. α-synuclein locus duplication as a cause of familial Parkinson's disease. *Lancet* **2004**, *364*, 1167–1169. [CrossRef]
14. Ibáñez, P.; Bonnet, A.-M.; Débarges, B.; Lohmann, E.; Tison, F.; Pollak, P.; Agid, Y.; Dürr, A.; Brice, A. Causal relation between alpha-synuclein gene duplication and familial Parkinson's disease. *Lancet (Lond. Engl.)* **2004**, *364*, 1169–1171. [CrossRef]
15. Nalls, M.A.; Blauwendraat, C.; Vallerga, C.L.; Heilbron, K.; Bandres-Ciga, S.; Chang, D.; Tan, M.; Kia, D.A.; Noyce, A.J.; Xue, A.; et al. Identification of novel risk loci, causal insights, and heritable risk for Parkinson's disease: A meta-analysis of genome-wide association studies. *Lancet Neurol.* **2019**, *18*, 1091–1102. [CrossRef]
16. Satake, W.; Nakabayashi, Y.; Mizuta, I.; Hirota, Y.; Ito, C.; Kubo, M.; Kawaguchi, T.; Tsunoda, T.; Watanabe, M.; Takeda, A.; et al. Genome-wide association study identifies common variants at four loci as genetic risk factors for Parkinson's disease. *Nat. Genet.* **2009**, *41*, 1303–1307. [CrossRef]
17. Simón-Sánchez, J.; Schulte, C.; Bras, J.M.; Sharma, M.; Gibbs, J.R.; Berg, D.; Paisan-Ruiz, C.; Lichtner, P.; Scholz, S.W.; Hernandez, D.G.; et al. Genome-wide association study reveals genetic risk underlying Parkinson's disease. *Nat. Genet.* **2009**, *41*, 1308–1312. [CrossRef]
18. Marchese, D.; Botta-Orfila, T.; Cirillo, D.; Rodriguez, J.A.; Livi, C.M.; Fernández-Santiago, R.; Ezquerra, M.; Martí, M.J.; Bechara, E.; Tartaglia, G.G. Discovering the 3′ UTR-mediated regulation of alpha-synuclein. *Nucleic Acids Res.* **2017**, *45*, 12888–12903. [CrossRef]
19. Tseng, E.; Rowell, W.J.; Glenn, O.-C.; Hon, T.; Barrera, J.; Kujawa, S.; Chiba-Falek, O. The Landscape of SNCA Transcripts Across Synucleinopathies: New Insights From Long Reads Sequencing Analysis. *Front. Genet.* **2019**, *10*, 584. [CrossRef]
20. Langmyhr, M.; Henriksen, S.P.; Cappelletti, C.; van de Berg, W.D.J.; Pihlstrøm, L.; Toft, M. Allele-specific expression of Parkinson's disease susceptibility genes in human brain. *Sci. Rep.* **2021**, *11*, 504. [CrossRef]
21. Guhathakurta, S.; Kim, J.; Adams, L.; Basu, S.; Song, M.K.; Adler, E.; Je, G.; Fiadeiro, M.B.; Kim, Y. Targeted attenuation of elevated histone marks at SNCA alleviates α-synuclein in Parkinson's disease. *EMBO Mol. Med.* **2021**, *13*, e12188. [CrossRef]
22. Pankratz, N.; Dumitriu, A.; Hetrick, K.N.; Sun, M.; Latourelle, J.C.; Wilk, J.B.; Halter, C.; Doheny, K.F.; Gusella, J.F.; Nichols, W.C.; et al. Copy Number Variation in Familial Parkinson Disease. *PLoS ONE* **2011**, *6*, e20988. [CrossRef]
23. Perez-Rodriguez, D.; Kalyva, M.; Leija-Salazar, M.; Lashley, T.; Tarabichi, M.; Chelban, V.; Gentleman, S.; Schottlaender, L.; Franklin, H.; Vasmatzis, G.; et al. Investigation of somatic CNVs in brains of synucleinopathy cases using targeted SNCA analysis and single cell sequencing. *Acta Neuropathol. Commun.* **2019**, *7*, 219. [CrossRef]

24. Mokretar, K.; Pease, D.; Taanman, J.-W.; Soenmez, A.; Ejaz, A.; Lashley, T.; Ling, H.; Gentleman, S.; Houlden, H.; Holton, J.L.; et al. Somatic copy number gains of α-synuclein (SNCA) in Parkinson's disease and multiple system atrophy brains. *Brain* **2018**, *141*, 2419–2431. [CrossRef]
25. Edwards, T.L.; Scott, W.K.; Almonte, C.; Burt, A.; Powell, E.H.; Beecham, G.W.; Wang, L.; Züchner, S.; Konidari, I.; Wang, G.; et al. Genome-wide association study confirms SNPs in SNCA and the MAPT region as common risk factors for Parkinson disease. *Ann. Hum. Genet.* **2010**, *74*, 97–109. [CrossRef]
26. Lesage, S.; Houot, M.; Mangone, G.; Tesson, C.; Bertrand, H.; Forlani, S.; Anheim, M.; Brefel-Courbon, C.; Broussolle, E.; Thobois, S.; et al. Genetic and Phenotypic Basis of Autosomal Dominant Parkinson's Disease in a Large Multi-Center Cohort. *Front. Neurol.* **2020**, *11*, 682. [CrossRef] [PubMed]
27. Schulz-Schaeffer, W.J. The synaptic pathology of α-synuclein aggregation in dementia with Lewy bodies, Parkinson's disease and Parkinson's disease dementia. *Acta Neuropathol.* **2010**, *120*, 131–143. [CrossRef]
28. Moussaud, S.; Jones, D.R.; Moussaud-Lamodière, E.L.; Delenclos, M.; Ross, O.A.; McLean, P.J. Alpha-synuclein and tau: Teammates in neurodegeneration? *Mol. Neurodegener.* **2014**, *9*, 43. [CrossRef] [PubMed]
29. Ingelsson, M. Alpha-Synuclein Oligomers—Neurotoxic Molecules in Parkinson's Disease and Other Lewy Body Disorders. *Front. Neurosci.* **2016**, *10*, 408. [CrossRef] [PubMed]
30. Theillet, F.-X.; Binolfi, A.; Bekei, B.; Martorana, A.; Rose, H.M.; Stuiver, M.; Verzini, S.; Lorenz, D.; van Rossum, M.; Goldfarb, D.; et al. Structural disorder of monomeric α-synuclein persists in mammalian cells. *Nature* **2016**, *530*, 45–50. [CrossRef]
31. Bartels, T.; Choi, J.G.; Selkoe, D.J. α-Synuclein occurs physiologically as a helically folded tetramer that resists aggregation. *Nature* **2011**, *477*, 107–110. [CrossRef]
32. Volpicelli-Daley, L.; Brundin, P. Prion-like propagation of pathology in Parkinson disease. *Handb. Clin. Neurol.* **2018**, *153*, 321–335. [CrossRef] [PubMed]
33. Brás, I.C.; Outeiro, T.F. Alpha-Synuclein: Mechanisms of Release and Pathology Progression in Synucleinopathies. *Cells* **2021**, *10*, 375. [CrossRef]
34. Massey, A.R.; Monogue, B.; Chen, Y.; Lesteberg, K.; Johnson, M.E.; Bergkvist, L.; Steiner, J.A.; Ma, J.; Mahalingam, R.; Kleinschmidt-Demasters, B.K.; et al. Alpha-synuclein supports interferon stimulated gene expression in neurons. *bioRxiv* **2020**. [CrossRef]
35. Burré, J. The synaptic function of α-synuclein. *J. Parkinson's Dis.* **2015**, *5*, 699–713. [CrossRef]
36. Burré, J.; Sharma, M.; Südhof, T.C. Cell Biology and Pathophysiology of α-Synuclein. *Cold Spring Harb. Perspect. Med.* **2018**, *8*, a024091. [CrossRef] [PubMed]
37. Bonini, N.M.; Giasson, B.I. Snaring the Function of α-Synuclein. *Cell* **2005**, *123*, 359–361. [CrossRef] [PubMed]
38. Man, W.K.; Tahirbegi, B.; Vrettas, M.D.; Preet, S.; Ying, L.; Vendruscolo, M.; De Simone, A.; Fusco, G. The docking of synaptic vesicles on the presynaptic membrane induced by α-synuclein is modulated by lipid composition. *Nat. Commun.* **2021**, *12*, 927. [CrossRef]
39. Maroteaux, L.; Campanelli, J.; Scheller, R. Synuclein: A neuron-specific protein localized to the nucleus and presynaptic nerve terminal. *J. Neurosci.* **1988**, *8*, 2804–2815. [CrossRef]
40. Spillantini, M.G.; Lee, V.M.-Y.; Trojanowski, J.Q.; Jakes, R.; Goedert, M. alpha-Synuclein in Lewy bodies. *Nature* **1997**, *388*, 839–840. [CrossRef]
41. Shahmoradian, S.H.; Lewis, A.J.; Genoud, C.; Hench, J.; Moors, T.E.; Navarro, P.P.; Castaño-Díez, D.; Schweighauser, G.; Graff-Meyer, A.; Goldie, K.N.; et al. Lewy pathology in Parkinson's disease consists of crowded organelles and lipid membranes. *Nat. Neurosci.* **2019**, *22*, 1099–1109. [CrossRef]
42. Bussell, R.; Eliezer, D. A Structural and Functional Role for 11-mer Repeats in α-Synuclein and Other Exchangeable Lipid Binding Proteins. *J. Mol. Biol.* **2003**, *329*, 763–778. [CrossRef]
43. Musteikytė, G.; Jayaram, A.K.; Xu, C.K.; Vendruscolo, M.; Krainer, G.; Knowles, T.P.J. Interactions of α-synuclein oligomers with lipid membranes. *Biochim. Biophys. Acta (BBA)-Biomembr.* **2021**, *1863*, 183536. [CrossRef] [PubMed]
44. Adão, R.; Cruz, P.F.; Vaz, D.C.; Fonseca, F.; Pedersen, J.N.; Ferreira-da-Silva, F.; Brito, R.M.M.; Ramos, C.H.I.; Otzen, D.; Keller, S.; et al. DIBMA nanodiscs keep α-synuclein folded. *Biochim. Biophys. Acta (BBA)-Biomembr.* **2020**, *1862*, 183314. [CrossRef] [PubMed]
45. Jo, E.; Fuller, N.; Rand, R.P.; St, George-Hyslop, P.; Fraser, P.E. Defective membrane interactions of familial Parkinson's disease mutant A30P α-synuclein. *J. Mol. Biol.* **2002**, *315*, 799–807. [CrossRef] [PubMed]
46. Raben, D.M.; Barber, C.N. Phosphatidic acid and neurotransmission. *Adv. Biol. Regul.* **2017**, *63*, 15–21. [CrossRef]
47. Fantini, J.; Carlus, D.; Yahi, N. The fusogenic tilted peptide (67–78) of α-synuclein is a cholesterol binding domain. *Biochim. Biophys. Acta (BBA)-Biomembr.* **2011**, *1808*, 2343–2351. [CrossRef]
48. Mahfoud, R.; Garmy, N.; Maresca, M.; Yahi, N.; Puigserver, A.; Fantini, J. Identification of a Common Sphingolipid-binding Domain in Alzheimer, Prion, and HIV-1 Proteins. *J. Biol. Chem.* **2002**, *277*, 11292–11296. [CrossRef]
49. Bartels, T.; Ahlstrom, L.S.; Leftin, A.; Kamp, F.; Haass, C.; Brown, M.F.; Beyer, K. The N-Terminus of the Intrinsically Disordered Protein α-Synuclein Triggers Membrane Binding and Helix Folding. *Biophys. J.* **2010**, *99*, 2116–2124. [CrossRef]
50. Davidson, W.S.; Jonas, A.; Clayton, D.F.; George, J.M. Stabilization of α-Synuclein Secondary Structure upon Binding to Synthetic Membranes. *J. Biol. Chem.* **1998**, *273*, 9443–9449. [CrossRef]
51. Ferreon, A.C.M.; Gambin, Y.; Lemke, E.A.; Deniz, A.A. Interplay of α-synuclein binding and conformational switching probed by single-molecule fluorescence. *Proc. Natl. Acad. Sci. USA* **2009**, *106*, 5645–5650. [CrossRef]

52. Meade, R.M.; Fairlie, D.P.; Mason, J.M. Alpha-synuclein structure and Parkinson's disease–lessons and emerging principles. *Mol. Neurodegener.* **2019**, *14*, 29. [CrossRef]
53. Bisaglia, M.; Tessari, I.; Pinato, L.; Bellanda, M.; Giraudo, S.; Fasano, M.; Bergantino, E.; Bubacco, L.; Mammi, S. A Topological Model of the Interaction between α-Synuclein and Sodium Dodecyl Sulfate Micelles. *Biochemistry* **2005**, *44*, 329–339. [CrossRef] [PubMed]
54. Fusco, G.; De Simone, A.; Gopinath, T.; Vostrikov, V.; Vendruscolo, M.; Dobson, C.M.; Veglia, G. Direct observation of the three regions in α-synuclein that determine its membrane-bound behaviour. *Nat. Commun.* **2014**, *5*, 3827. [CrossRef]
55. Eliezer, D.; Kutluay, E.; Bussell, R.; Browne, G. Conformational Properties of a-Synuclein in its Free and Lipid-associated States. *J. Mol. Biol.* **2001**, *307*, 1061–1073. [CrossRef]
56. Zhang, J.; Li, X.; Li, J.-D. The Roles of Post-translational Modifications on α-Synuclein in the Pathogenesis of Parkinson's Diseases. *Front. Neurosci.* **2019**, *13*, 381. [CrossRef] [PubMed]
57. Watson, H. Biological membranes. *Essays Biochem.* **2015**, *59*, 43–69. [CrossRef] [PubMed]
58. Ingólfsson, H.I.; Carpenter, T.S.; Bhatia, H.; Bremer, P.-T.; Marrink, S.J.; Lightstone, F.C. Computational Lipidomics of the Neuronal Plasma Membrane. *Biophys. J.* **2017**, *113*, 2271–2280. [CrossRef]
59. Chiricozzi, E.; Lunghi, G.; Di Biase, E.; Fazzari, M.; Sonnino, S.; Mauri, L. GM1 Ganglioside Is A Key Factor in Maintaining the Mammalian Neuronal Functions Avoiding Neurodegeneration. *Int. J. Mol. Sci.* **2020**, *21*, 868. [CrossRef]
60. Seyfried, T.N.; Choi, H.; Chevalier, A.; Hogan, D.; Akgoc, Z.; Schneider, J.S. Sex-Related Abnormalities in Substantia Nigra Lipids in Parkinson's Disease. *ASN Neuro* **2018**, *10*, 175909141878188. [CrossRef]
61. Fusco, G.; Pape, T.; Stephens, A.D.; Mahou, P.; Costa, A.R.; Kaminski, C.F.; Kaminski Schierle, G.S.; Vendruscolo, M.; Veglia, G.; Dobson, C.M.; et al. Structural basis of synaptic vesicle assembly promoted by α-synuclein. *Nat. Commun.* **2016**, *7*, 12563. [CrossRef]
62. Man, W.K.; De Simone, A.; Barritt, J.D.; Vendruscolo, M.; Dobson, C.M.; Fusco, G. A Role of Cholesterol in Modulating the Binding of α-Synuclein to Synaptic-Like Vesicles. *Front. Neurosci.* **2020**, *14*, 18. [CrossRef]
63. Fanning, S.; Haque, A.; Imberdis, T.; Baru, V.; Barrasa, M.I.; Nuber, S.; Termine, D.; Ramalingam, N.; Ho, G.P.H.; Noble, T.; et al. Lipidomic Analysis of α-Synuclein Neurotoxicity Identifies Stearoyl CoA Desaturase as a Target for Parkinson Treatment. *Mol. Cell* **2019**, *73*, 1001–1014.e8. [CrossRef]
64. Xicoy, H.; Brouwers, J.F.; Wieringa, B.; Martens, G.J.M. Explorative Combined Lipid and Transcriptomic Profiling of Substantia Nigra and Putamen in Parkinson's Disease. *Cells* **2020**, *9*, 1966. [CrossRef]
65. Fernández-Irigoyen, J.; Cartas-Cejudo, P.; Iruarrizaga-Lejarreta, M.; Santamaría, E. Alteration in the Cerebrospinal Fluid Lipidome in Parkinson's Disease: A Post-Mortem Pilot Study. *Biomedicines* **2021**, *9*, 491. [CrossRef]
66. Lv, Z.; Hashemi, M.; Banerjee, S.; Zagorski, K.; Rochet, J.-C.; Lyubchenko, Y.L. Assembly of α-synuclein aggregates on phospholipid bilayers. *Biochim. Biophys. Acta-Proteins Proteom.* **2019**, *1867*, 802–812. [CrossRef]
67. Hattingen, E.; Magerkurth, J.; Pilatus, U.; Mozer, A.; Seifried, C.; Steinmetz, H.; Zanella, F.; Hilker, R. Phosphorus and proton magnetic resonance spectroscopy demonstrates mitochondrial dysfunction in early and advanced Parkinson's disease. *Brain* **2009**, *132*, 3285–3297. [CrossRef]
68. Wang, S.; Zhang, S.; Liou, L.-C.; Ren, Q.; Zhang, Z.; Caldwell, G.A.; Caldwell, K.A.; Witt, S.N. Phosphatidylethanolamine deficiency disrupts α-synuclein homeostasis in yeast and worm models of Parkinson disease. *Proc. Natl. Acad. Sci. USA* **2014**, *111*, E3976–E3985. [CrossRef]
69. Lou, X.; Kim, J.; Hawk, B.J.; Shin, Y.-K. α-Synuclein may cross-bridge v-SNARE and acidic phospholipids to facilitate SNARE-dependent vesicle docking. *Biochem. J.* **2017**, *474*, 2039–2049. [CrossRef]
70. Abbott, S.K.; Li, H.; Muñoz, S.S.; Knoch, B.; Batterham, M.; Murphy, K.E.; Halliday, G.M.; Garner, B. Altered ceramide acyl chain length and ceramide synthase gene expression in Parkinson's disease. *Mov. Disord.* **2014**, *29*, 518–526. [CrossRef]
71. Kim, W.S.; Halliday, G.M. Changes in sphingomyelin level affect alpha-synuclein and ABCA5 expression. *J. Parkinson's Dis.* **2012**, *2*, 41–46. [CrossRef]
72. Mesa-Herrera, F.; Taoro-González, L.; Valdés-Baizabal, V.-B.; Diaz, M.; Marín, R. Lipid and Lipid Raft Alteration in Aging and Neurodegenerative Diseases: A Window for the Development of New Biomarkers. *Int. J. Mol. Sci.* **2019**, *20*, 3810. [CrossRef]
73. Martinez, Z.; Zhu, M.; Han, S.; Fink, A.L. GM1 Specifically Interacts with α-Synuclein and Inhibits Fibrillation. *Biochemistry* **2007**, *46*, 1868–1877. [CrossRef]
74. Grey, M.; Dunning, C.J.; Gaspar, R.; Grey, C.; Brundin, P.; Sparr, E.; Linse, S. Acceleration of α-Synuclein Aggregation by Exosomes. *J. Biol. Chem.* **2015**, *290*, 2969–2982. [CrossRef]
75. Lee, Y.J.; Wang, S.; Slone, S.R.; Yacoubian, T.A.; Witt, S.N. Defects in Very Long Chain Fatty Acid Synthesis Enhance Alpha-Synuclein Toxicity in a Yeast Model of Parkinson's Disease. *PLoS ONE* **2011**, *6*, e15946. [CrossRef]
76. Appel-Cresswell, S.; Vilarino-Guell, C.; Encarnacion, M.; Sherman, H.; Yu, I.; Shah, B.; Weir, D.; Thompson, C.; Szu-Tu, C.; Trinh, J.; et al. Alpha-synuclein p.H50Q, a novel pathogenic mutation for Parkinson's disease. *Mov. Disord.* **2013**, *28*, 811–813. [CrossRef] [PubMed]
77. Ramakrishnan, M.; Jensen, P.H.; Marsh, D. α-Synuclein Association with Phosphatidylglycerol Probed by Lipid Spin Labels. *Biochemistry* **2003**, *42*, 12919–12926. [CrossRef] [PubMed]
78. Schommer, J.; Marwarha, G.; Nagamoto-Combs, K.; Ghribi, O. Palmitic Acid-Enriched Diet Increases α-Synuclein and Tyrosine Hydroxylase Expression Levels in the Mouse Brain. *Front. Neurosci.* **2018**, *12*, 552. [CrossRef] [PubMed]

79. Sharon, R.; Bar-Joseph, I.; Frosch, M.P.; Walsh, D.M.; Hamilton, J.A.; Selkoe, D.J. The Formation of Highly Soluble Oligomers of α-Synuclein Is Regulated by Fatty Acids and Enhanced in Parkinson's Disease. *Neuron* **2003**, *37*, 583–595. [CrossRef]
80. Vincent, B.M.; Tardiff, D.F.; Piotrowski, J.S.; Aron, R.; Lucas, M.C.; Chung, C.Y.; Bacherman, H.; Chen, Y.; Pires, M.; Subramaniam R.; et al. Inhibiting Stearoyl-CoA Desaturase Ameliorates α-Synuclein Cytotoxicity. *Cell Rep.* **2018**, *25*, 2742–2754.e31. [CrossRef] [PubMed]
81. Marin, R.; Fabelo, N.; Martín, V.; Garcia-Esparcia, P.; Ferrer, I.; Quinto-Alemany, D.; Díaz, M. Anomalies occurring in lipid profiles and protein distribution in frontal cortex lipid rafts in dementia with Lewy bodies disclose neurochemical traits partially shared by Alzheimer's and Parkinson's diseases. *Neurobiol. Aging* **2017**, *49*, 52–59. [CrossRef]
82. Yakunin, E.; Loeb, V.; Kisos, H.; Biala, Y.; Yehuda, S.; Yaari, Y.; Selkoe, D.J.; Sharon, R. α-Synuclein Neuropathology is Controlled by Nuclear Hormone Receptors and Enhanced by Docosahexaenoic Acid in A Mouse Model for Parkinson's Disease. *Brain Pathol.* **2012**, *22*, 280–294. [CrossRef] [PubMed]
83. Galvagnion, C.; Brown, J.W.P.; Ouberai, M.M.; Flagmeier, P.; Vendruscolo, M.; Buell, A.K.; Sparr, E.; Dobson, C.M. Chemical properties of lipids strongly affect the kinetics of the membrane-induced aggregation of α-synuclein. *Proc. Natl. Acad. Sci. USA* **2016**, *113*, 7065–7070. [CrossRef]
84. Pyszko, J.; Strosznajder, J.B. Sphingosine Kinase 1 and Sphingosine-1-Phosphate in Oxidative Stress Evoked by 1-Methyl-4-Phenylpyridinium (MPP+) in Human Dopaminergic Neuronal Cells. *Mol. Neurobiol.* **2014**, *50*, 38–48. [CrossRef]
85. Alcalay, R.N.; Mallett, V.; Vanderperre, B.; Tavassoly, O.; Dauvilliers, Y.; Wu, R.Y.J.; Ruskey, J.A.; Leblond, C.S.; Ambalavanan, A.; Laurent, S.B.; et al. SMPD1 mutations, activity, and α-synuclein accumulation in Parkinson's disease. *Mov. Disord.* **2019**, *34*, 526–535. [CrossRef]
86. Pyszko, J.A.; Strosznajder, J.B. Original article The key role of sphingosine kinases in the molecular mechanism of neuronal cell survival and death in an experimental model of Parkinson's disease. *Folia Neuropathol.* **2014**, *52*, 260–269. [CrossRef]
87. Bae, E.-J.; Lee, H.-J.; Jang, Y.-H.; Michael, S.; Masliah, E.; Min, D.S.; Lee, S.-J. Phospholipase D1 regulates autophagic flux and clearance of α-synuclein aggregates. *Cell Death Differ.* **2014**, *21*, 1132–1141. [CrossRef]
88. Sidransky, E.; Lopez, G. The link between the GBA gene and parkinsonism. *Lancet Neurol.* **2012**, *11*, 986–998. [CrossRef]
89. Rocha, E.M.; Smith, G.A.; Park, E.; Cao, H.; Brown, E.; Hallett, P.; Isacson, O. Progressive decline of glucocerebrosidase in aging and Parkinson's disease. *Ann. Clin. Transl. Neurol.* **2015**, *2*, 433–438. [CrossRef] [PubMed]
90. Xicoy, H.; Peñuelas, N.; Vila, M.; Laguna, A. Autophagic- and Lysosomal-Related Biomarkers for Parkinson's Disease: Lights and Shadows. *Cells* **2019**, *8*, 1317. [CrossRef]
91. Belarbi, K.; Cuvelier, E.; Bonte, M.-A.; Desplanque, M.; Gressier, B.; Devos, D.; Chartier-Harlin, M.-C. Glycosphingolipids and neuroinflammation in Parkinson's disease. *Mol. Neurodegener.* **2020**, *15*, 59. [CrossRef]
92. Van Dijk, K.D.; Persichetti, E.; Chiasserini, D.; Eusebi, P.; Beccari, T.; Calabresi, P.; Berendse, H.W.; Parnetti, L.; van de Berg, W.D.J. Changes in endolysosomal enzyme activities in cerebrospinal fluid of patients with Parkinson's disease. *Mov. Disord.* **2013**, *28*, 747–754. [CrossRef]
93. McGlinchey, R.P.; Lee, J.C. Cysteine cathepsins are essential in lysosomal degradation of α-synuclein. *Proc. Natl. Acad. Sci. USA* **2015**, *112*, 9322–9327. [CrossRef]
94. Brekk, O.R.; Korecka, J.A.; Crapart, C.C.; Huebecker, M.; MacBain, Z.K.; Rosenthal, S.A.; Sena-Esteves, M.; Priestman, D.A.; Platt, F.M.; Isacson, O.; et al. Upregulating β-hexosaminidase activity in rodents prevents α-synuclein lipid associations and protects dopaminergic neurons from α-synuclein-mediated neurotoxicity. *Acta Neuropathol. Commun.* **2020**, *8*, 127. [CrossRef]
95. Pike, L.J. Lipid rafts: Bringing order to chaos. *J. Lipid Res.* **2003**, *44*, 655–667. [CrossRef]
96. Zabrocki, P.; Bastiaens, I.; Delay, C.; Bammens, T.; Ghillebert, R.; Pellens, K.; De Virgilio, C.; Van Leuven, F.; Winderickx, J. Phosphorylation, lipid raft interaction and traffic of α-synuclein in a yeast model for Parkinson. *Biochim. Biophys. Acta* **2008**, *1783*, 1767–1780. [CrossRef]
97. Fortin, D.L.; Troyer, M.D.; Nakamura, K.; Kubo, S.I.; Anthony, M.D.; Edwards, R.H. Lipid Rafts Mediate the Synaptic Localization of -Synuclein. *J. Neurosci.* **2004**, *24*, 6715–6723. [CrossRef]
98. Perissinotto, F.; Stani, C.; De Cecco, E.; Vaccari, L.; Rondelli, V.; Posocco, P.; Parisse, P.; Scaini, D.; Legname, G.; Casalis, L. Iron-mediated interaction of alpha synuclein with lipid raft model membranes. *Nanoscale* **2020**, *12*, 7631–7640. [CrossRef]
99. Varkey, J.; Isas, J.M.; Mizuno, N.; Jensen, M.B.; Bhatia, V.K.; Jao, C.C.; Petrlova, J.; Voss, J.C.; Stamou, D.G.; Steven, A.C.; et al. Membrane Curvature Induction and Tubulation Are Common Features of Synucleins and Apolipoproteins. *J. Biol. Chem.* **2010**, *285*, 32486–32493. [CrossRef] [PubMed]
100. Pandey, A.P.; Haque, F.; Rochet, J.-C.; Hovis, J.S. α-Synuclein-Induced Tubule Formation in Lipid Bilayers. *J. Phys. Chem. B* **2011**, *115*, 5886–5893. [CrossRef] [PubMed]
101. Westphal, C.H.; Chandra, S.S. Monomeric Synucleins Generate Membrane Curvature. *J. Biol. Chem.* **2013**, *288*, 1829–1840. [CrossRef]
102. Shen, H.; Pirruccello, M.; De Camilli, P. SnapShot: Membrane Curvature Sensors and Generators. *Cell* **2012**, *150*, 1300–1300.e2. [CrossRef]
103. Wang, H.-L.; Lu, C.-S.; Yeh, T.-H.; Shen, Y.-M.; Weng, Y.-H.; Huang, Y.-Z.; Chen, R.-S.; Liu, Y.-C.; Cheng, Y.-C.; Chang, H.-C.; et al. Combined Assessment of Serum Alpha-Synuclein and Rab35 is a Better Biomarker for Parkinson's Disease. *J. Clin. Neurol.* **2019**, *15*, 488–495. [CrossRef]

104. Schechter, M.; Atias, M.; Abd Elhadi, S.; Davidi, D.; Gitler, D.; Sharon, R. α-Synuclein facilitates endocytosis by elevating the steady-state levels of phosphatidylinositol 4,5-bisphosphate. *J. Biol. Chem.* **2020**, *295*, 18076–18090. [CrossRef]
105. Tosatto, L.; Andrighetti, A.O.; Plotegher, N.; Antonini, V.; Tessari, I.; Ricci, L.; Bubacco, L.; Dalla Serra, M. Alpha-synuclein pore forming activity upon membrane association. *Biochim. Biophys. Acta* **2012**, *1818*, 2876–2883. [CrossRef] [PubMed]
106. Takamori, S.; Holt, M.; Stenius, K.; Lemke, E.A.; Grønborg, M.; Riedel, D.; Urlaub, H.; Schenck, S.; Brügger, B.; Ringler, P.; et al. Molecular Anatomy of a Trafficking Organelle. *Cell* **2006**, *127*, 831–846. [CrossRef] [PubMed]
107. Middleton, E.R.; Rhoades, E. Effects of curvature and composition on α-synuclein binding to lipid vesicles. *Biophys. J.* **2010**, *99*, 2279–2288. [CrossRef] [PubMed]
108. Runfola, M.; De Simone, A.; Vendruscolo, M.; Dobson, C.M.; Fusco, G. The N-terminal Acetylation of α-Synuclein Changes the Affinity for Lipid Membranes but not the Structural Properties of the Bound State. *Sci. Rep.* **2020**, *10*, 204. [CrossRef] [PubMed]
109. Maltsev, A.S.; Ying, J.; Bax, A. Impact of N-Terminal Acetylation of α-Synuclein on Its Random Coil and Lipid Binding Properties. *Biochemistry* **2012**, *51*, 5004–5013. [CrossRef] [PubMed]
110. O'Leary, E.I.; Jiang, Z.; Strub, M.-P.; Lee, J.C. Effects of phosphatidylcholine membrane fluidity on the conformation and aggregation of N-terminally acetylated α-synuclein. *J. Biol. Chem.* **2018**, *293*, 11195–11205. [CrossRef] [PubMed]
111. Samuel, F.; Flavin, W.P.; Iqbal, S.; Pacelli, C.; Sri Renganathan, S.D.; Trudeau, L.-E.; Campbell, E.M.; Fraser, P.E.; Tandon, A. Effects of Serine 129 Phosphorylation on α-Synuclein Aggregation, Membrane Association, and Internalization. *J. Biol. Chem.* **2016**, *291*, 4374–4385. [CrossRef]
112. Dikiy, I.; Fauvet, B.; Jovičić, A.; Mahul-Mellier, A.-L.; Desobry, C.; El-Turk, F.; Gitler, A.D.; Lashuel, H.A.; Eliezer, D. Semisynthetic and in Vitro Phosphorylation of Alpha-Synuclein at Y39 Promotes Functional Partly Helical Membrane-Bound States Resembling Those Induced by PD Mutations. *ACS Chem. Biol.* **2016**, *11*, 2428–2437. [CrossRef] [PubMed]
113. Lindau, M.; Alvarez de Toledo, G. The fusion pore. *Biochim. Biophys. Acta* **2003**, *1641*, 167–173. [CrossRef]
114. Pevsner, J.; Hsu, S.-C.; Braun, J.E.A.; Calakos, N.; Ting, A.E.; Bennett, M.K.; Scheller, R.H. Specificity and regulation of a synaptic vesicle docking complex. *Neuron* **1994**, *13*, 353–361. [CrossRef]
115. Bost, A.; Shaib, A.H.; Schwarz, Y.; Niemeyer, B.A.; Becherer, U. Large dense-core vesicle exocytosis from mouse dorsal root ganglion neurons is regulated by neuropeptide Y. *Neuroscience* **2017**, *346*, 1–13. [CrossRef]
116. Staal, R.G.W.; Mosharov, E.V.; Sulzer, D. Dopamine neurons release transmitter via a flickering fusion pore. *Nat. Neurosci.* **2004**, *7*, 341–346. [CrossRef]
117. Arispe, N.; Pollard, H.B.; Rojas, E. Giant multilevel cation channels formed by Alzheimer disease amyloid beta-protein [A beta P-(1-40)] in bilayer membranes. *Proc. Natl. Acad. Sci. USA* **1993**, *90*, 10573–10577. [CrossRef]
118. Bode, D.C.; Baker, M.D.; Viles, J.H. Ion Channel Formation by Amyloid-β42 Oligomers but Not Amyloid-β40 in Cellular Membranes. *J. Biol. Chem.* **2017**, *292*, 1404–1413. [CrossRef]
119. Wolozin, B. Cholesterol and the biology of Alzheimer's disease. *Neuron* **2004**, *41*, 7–10. [CrossRef]
120. Williams, T.L.; Serpell, L.C. Membrane and surface interactions of Alzheimer's Aβ peptide—Insights into the mechanism of cytotoxicity. *FEBS J.* **2011**, *278*, 3905–3917. [CrossRef] [PubMed]
121. Arispe, N.; Doh, M. Plasma membrane cholesterol controls the cytotoxicity of Alzheimer's disease AbetaP (1-40) and (1-42) peptides. *FASEB J.* **2002**, *16*, 1526–1536. [CrossRef]
122. Hong, S.; Ostaszewski, B.L.; Yang, T.; O'Malley, T.T.; Jin, M.; Yanagisawa, K.; Li, S.; Bartels, T.; Selkoe, D.J. Soluble Aβ Oligomers Are Rapidly Sequestered from Brain ISF In Vivo and Bind GM1 Ganglioside on Cellular Membranes. *Neuron* **2014**, *82*, 308–319. [CrossRef] [PubMed]
123. Lashuel, H.A.; Petre, B.M.; Wall, J.; Simon, M.; Nowak, R.J.; Walz, T.; Lansbury, P.T. α-Synuclein, Especially the Parkinson's Disease-associated Mutants, Forms Pore-like Annular and Tubular Protofibrils. *J. Mol. Biol.* **2002**, *322*, 1089–1102. [CrossRef]
124. Tsigelny, I.F.; Sharikov, Y.; Wrasidlo, W.; Gonzalez, T.; Desplats, P.A.; Crews, L.; Spencer, B.; Masliah, E. Role of α-synuclein penetration into the membrane in the mechanisms of oligomer pore formation. *FEBS J.* **2012**, *279*, 1000–1013. [CrossRef]
125. Logan, T.; Bendor, J.; Toupin, C.; Thorn, K.; Edwards, R.H. α-Synuclein promotes dilation of the exocytotic fusion pore. *Nat. Neurosci.* **2017**, *20*, 681–689. [CrossRef]
126. Runwal, G.; Edwards, R.H. The Membrane Interactions of Synuclein: Physiology and Pathology. *Annu. Rev. Pathol. Mech. Dis.* **2021**, *16*, 465–485. [CrossRef] [PubMed]
127. Abbineni, P.S.; Bohannon, K.P.; Bittner, M.A.; Axelrod, D.; Holz, R.W. Identification of β-synuclein on secretory granules in chromaffin cells and the effects of α- and β-synuclein on post-fusion BDNF discharge and fusion pore expansion. *Neurosci. Lett.* **2019**, *699*, 134–139. [CrossRef]
128. Larsen, K.E.; Schmitz, Y.; Troyer, M.D.; Mosharov, E.; Dietrich, P.; Quazi, A.Z.; Savalle, M.; Nemani, V.; Chaudhry, F.A.; Edwards, R.H.; et al. α-Synuclein Overexpression in PC12 and Chromaffin Cells Impairs Catecholamine Release by Interfering with a Late Step in Exocytosis. *J. Neurosci.* **2006**, *26*, 11915–11922. [CrossRef]
129. Nemani, V.M.; Lu, W.; Berge, V.; Nakamura, K.; Onoa, B.; Lee, M.K.; Chaudhry, F.A.; Nicoll, R.A.; Edwards, R.H. Increased Expression of α-Synuclein Reduces Neurotransmitter Release by Inhibiting Synaptic Vesicle Reclustering after Endocytosis. *Neuron* **2010**, *65*, 66–79. [CrossRef]
130. Dingjan, I.; Linders, P.T.A.; Verboogen, D.R.J.; Revelo, N.H.; Ter Beest, M.; van den Bogaart, G. Endosomal and Phagosomal SNAREs. *Physiol. Rev.* **2018**, *98*, 1465–1492. [CrossRef]

131. Sun, J.; Wang, L.; Bao, H.; Premi, S.; Das, U.; Chapman, E.R.; Roy, S. Functional cooperation of α-synuclein and VAMP2 in synaptic vesicle recycling. *Proc. Natl. Acad. Sci. USA* **2019**, *116*, 11113–11115. [CrossRef]
132. Burre, J.; Sharma, M.; Tsetsenis, T.; Buchman, V.; Etherton, M.R.; Sudhof, T.C. α-Synuclein Promotes SNARE-Complex Assembly in Vivo and in Vitro. *Science* **2010**, *329*, 1663–1667. [CrossRef]
133. Choi, M.-G.; Kim, M.J.; Kim, D.-G.; Yu, R.; Jang, Y.-N.; Oh, W.-J. Sequestration of synaptic proteins by alpha-synuclein aggregates leading to neurotoxicity is inhibited by small peptide. *PLoS ONE* **2018**, *13*, e0195339. [CrossRef]
134. Sharma, M.; Burré, J.; Südhof, T.C. CSPα promotes SNARE-complex assembly by chaperoning SNAP-25 during synaptic activity. *Nat. Cell Biol.* **2011**, *13*, 30–39. [CrossRef]
135. Roosen, D.A.; Blauwendraat, C.; Cookson, M.R.; Lewis, P.A. DNAJC proteins and pathways to parkinsonism. *FEBS J.* **2019**, *286*, 3080–3094. [CrossRef]
136. Zaltieri, M.; Grigoletto, J.; Longhena, F.; Navarria, L.; Favero, G.; Castrezzati, S.; Colivicchi, M.A.; Della Corte, L.; Rezzani, R.; Pizzi, M.; et al. α-synuclein and synapsin III cooperatively regulate synaptic function in dopamine neurons. *J. Cell Sci.* **2015**, *128*, 2231–2243. [CrossRef]
137. Hoffmann, C.; Sansevrino, R.; Morabito, G.; Logan, C.; Vabulas, R.M.; Ulusoy, A.; Ganzella, M.; Milovanovic, D. Synapsin Condensates Recruit alpha-Synuclein. *J. Mol. Biol.* **2021**, *433*, 166961. [CrossRef]
138. Brose, N. For Better or for Worse: Complexins Regulate SNARE Function and Vesicle Fusion. *Traffic* **2008**, *9*, 1403–1413. [CrossRef] [PubMed]
139. Gong, J.; Lai, Y.; Li, X.; Wang, M.; Leitz, J.; Hu, Y.; Zhang, Y.; Choi, U.B.; Cipriano, D.; Pfuetzner, R.A.; et al. C-terminal domain of mammalian complexin-1 localizes to highly curved membranes. *Proc. Natl. Acad. Sci. USA* **2016**, *113*, E7590–E7599. [CrossRef]
140. Liu, J.; Bu, B.; Crowe, M.; Li, D.; Diao, J.; Ji, B. Membrane packing defects in synaptic vesicles recruit complexin and synuclein. *Phys. Chem. Chem. Phys.* **2021**, *23*, 2117–2125. [CrossRef]
141. Chandra, S.; Fornai, F.; Kwon, H.-B.; Yazdani, U.; Atasoy, D.; Liu, X.; Hammer, R.E.; Battaglia, G.; German, D.C.; Castillo, P.E.; et al. Double-knockout mice for α- and β-synucleins: Effect on synaptic functions. *Proc. Natl. Acad. Sci. USA* **2004**, *101*, 14966–14971. [CrossRef] [PubMed]
142. Chandra, S.; Gallardo, G.; Fernández-Chacón, R.; Schlüter, O.M.; Südhof, T.C. α-Synuclein Cooperates with CSPα in Preventing Neurodegeneration. *Cell* **2005**, *123*, 383–396. [CrossRef] [PubMed]
143. Nakata, Y.; Yasuda, T.; Fukaya, M.; Yamamori, S.; Itakura, M.; Nihira, T.; Hayakawa, H.; Kawanami, A.; Kataoka, M.; Nagai, M.; et al. Accumulation of α-Synuclein Triggered by Presynaptic Dysfunction. *J. Neurosci.* **2012**, *32*, 17186–17196. [CrossRef]
144. Faustini, G.; Longhena, F.; Varanita, T.; Bubacco, L.; Pizzi, M.; Missale, C.; Benfenati, F.; Björklund, A.; Spano, P.; Bellucci, A. Synapsin III deficiency hampers α-synuclein aggregation, striatal synaptic damage and nigral cell loss in an AAV-based mouse model of Parkinson's disease. *Acta Neuropathol.* **2018**, *136*, 621–639. [CrossRef]
145. Longhena, F.; Faustini, G.; Varanita, T.; Zaltieri, M.; Porrini, V.; Tessari, I.; Poliani, P.L.; Missale, C.; Borroni, B.; Padovani, A.; et al. Synapsin III is a key component of α-synuclein fibrils in Lewy bodies of PD brains. *Brain Pathol.* **2018**, *28*, 875–888. [CrossRef]
146. Cabin, D.E.; Shimazu, K.; Murphy, D.; Cole, N.B.; Gottschalk, W.; McIlwain, K.L.; Orrison, B.; Chen, A.; Ellis, C.E.; Paylor, R.; et al. Synaptic Vesicle Depletion Correlates with Attenuated Synaptic Responses to Prolonged Repetitive Stimulation in Mice Lacking α-Synuclein. *J. Neurosci.* **2002**, *22*, 8797–8807. [CrossRef]
147. Senior, S.L.; Ninkina, N.; Deacon, R.; Bannerman, D.; Buchman, V.L.; Cragg, S.J.; Wade-Martins, R. Increased striatal dopamine release and hyperdopaminergic-like behaviour in mice lacking both alpha-synuclein and gamma-synuclein. *Eur. J. Neurosci.* **2008**, *27*, 947–957. [CrossRef]
148. Greten-Harrison, B.; Polydoro, M.; Morimoto-Tomita, M.; Diao, L.; Williams, A.M.; Nie, E.H.; Makani, S.; Tian, N.; Castillo, P.E.; Buchman, V.L.; et al. αβγ-Synuclein triple knockout mice reveal age-dependent neuronal dysfunction. *Proc. Natl. Acad. Sci. USA* **2010**, *107*, 19573–19578. [CrossRef]
149. Fountaine, T.M.; Venda, L.L.; Warrick, N.; Christian, H.C.; Brundin, P.; Channon, K.M.; Wade-Martins, R. The effect of α-synuclein knockdown on MPP+ toxicity in models of human neurons. *Eur. J. Neurosci.* **2008**, *28*, 2459–2473. [CrossRef]
150. Guo, J.T.; Chen, A.Q.; Kong, Q.; Zhu, H.; Ma, C.M.; Qin, C. Inhibition of Vesicular Monoamine Transporter-2 Activity in α-Synuclein Stably Transfected SH-SY5Y Cells. *Cell. Mol. Neurobiol.* **2008**, *28*, 35–47. [CrossRef]
151. Bu, M.; Farrer, M.J.; Khoshbouei, H. Dynamic control of the dopamine transporter in neurotransmission and homeostasis. *NPJ Park. Dis.* **2021**, *7*, 22. [CrossRef]
152. Wersinger, C.; Prou, D.; Vernier, P.; Sidhu, A. Modulation of dopamine transporter function by α-synuclein is altered by impairment of cell adhesion and by induction of oxidative stress. *FASEB J.* **2003**, *17*, 2151–2153. [CrossRef]
153. Lee, F.J.S.; Liu, F.; Pristupa, Z.B.; Niznik, H.B. Direct binding and functional coupling of α-synuclein to the dopamine transporters accelerate dopamine-induced apoptosis. *FASEB J.* **2001**, *15*, 916–926. [CrossRef]
154. Gaugler, M.N.; Genc, O.; Bobela, W.; Mohanna, S.; Ardah, M.T.; El-Agnaf, O.M.; Cantoni, M.; Bensadoun, J.-C.; Schneggenburger, R.; Knott, G.W.; et al. Nigrostriatal overabundance of α-synuclein leads to decreased vesicle density and deficits in dopamine release that correlate with reduced motor activity. *Acta Neuropathol.* **2012**, *123*, 653–669. [CrossRef]
155. Lundblad, M.; Decressac, M.; Mattsson, B.; Bjorklund, A. Impaired neurotransmission caused by overexpression of α-synuclein in nigral dopamine neurons. *Proc. Natl. Acad. Sci. USA* **2012**, *109*, 3213–3219. [CrossRef] [PubMed]

156. Somayaji, M.; Cataldi, S.; Choi, S.J.; Edwards, R.H.; Mosharov, E.V.; Sulzer, D. A dual role for α-synuclein in facilitation and depression of dopamine release from substantia nigra neurons in vivo. *Proc. Natl. Acad. Sci. USA* **2020**, *117*, 32701–32710. [CrossRef] [PubMed]
157. Fanning, S.; Selkoe, D.; Dettmer, U. Vesicle trafficking and lipid metabolism in synucleinopathy. *Acta Neuropathol.* **2021**, *141*, 491–510. [CrossRef]
158. Yavich, L.; Oksman, M.; Tanila, H.; Kerokoski, P.; Hiltunen, M.; van Groen, T.; Puoliväli, J.; Männistö, P.T.; García-Horsman, A.; MacDonald, E.; et al. Locomotor activity and evoked dopamine release are reduced in mice overexpressing A30P-mutated human α-synuclein. *Neurobiol. Dis.* **2005**, *20*, 303–313. [CrossRef]
159. Bellani, S.; Sousa, V.L.; Ronzitti, G.; Valtorta, F.; Meldolesi, J.; Chieregatti, E. The regulation of synaptic function by α-synuclein. *Commun. Integr. Biol.* **2010**, *3*, 106–109. [CrossRef]
160. Liu, S.; Ninan, I.; Antonova, I.; Battaglia, F.; Trinchese, F.; Narasanna, A.; Kolodilov, N.; Dauer, W.; Hawkins, R.D.; Arancio, O. α-Synuclein produces a long-lasting increase in neurotransmitter release. *EMBO J.* **2004**, *23*, 4506–4516. [CrossRef]
161. Alarcón-Arís, D.; Pavia-Collado, R.; Miquel-Rio, L.; Coppola-Segovia, V.; Ferrés-Coy, A.; Ruiz-Bronchal, E.; Galofré, M.; Paz, V.; Campa, L.; Revilla, R.; et al. Anti-α-synuclein ASO delivered to monoamine neurons prevents α-synuclein accumulation in a Parkinson's disease-like mouse model and in monkeys. *EBioMedicine* **2020**, *59*, 102944. [CrossRef]
162. Hill, E.; Gowers, R.; Richardson, M.J.E.; Wall, M.J. α-Synuclein Aggregates Increase the Conductance of Substantia Nigra Dopamine Neurons, an Effect Partly Reversed by the KATP Channel Inhibitor Glibenclamide. *Eneuro* **2021**, *8*, ENEURO.0330-20.2020. [CrossRef]
163. Soll, L.G.; Eisen, J.N.; Vargas, K.J.; Medeiros, A.T.; Hammar, K.M.; Morgan, J.R. α-Synuclein-112 Impairs Synaptic Vesicle Recycling Consistent With Its Enhanced Membrane Binding Properties. *Front. Cell Dev. Biol.* **2020**, *8*, 405. [CrossRef]
164. Busch, D.J.; Oliphint, P.A.; Walsh, R.B.; Banks, S.M.L.; Woods, W.S.; George, J.M.; Morgan, J.R. Acute increase of α-synuclein inhibits synaptic vesicle recycling evoked during intense stimulation. *Mol. Biol. Cell* **2014**, *25*, 3926–3941. [CrossRef]
165. Darios, F.; Ruipérez, V.; López, I.; Villanueva, J.; Gutierrez, L.M.; Davletov, B. α-Synuclein sequesters arachidonic acid to modulate SNARE-mediated exocytosis. *EMBO Rep.* **2010**, *11*, 528–533. [CrossRef]
166. Choi, B.-K.; Choi, M.-G.; Kim, J.-Y.; Yang, Y.; Lai, Y.; Kweon, D.-H.; Lee, N.K.; Shin, Y.-K. Large α-synuclein oligomers inhibit neuronal SNARE-mediated vesicle docking. *Proc. Natl. Acad. Sci. USA* **2013**, *110*, 4087–4092. [CrossRef]
167. Yoo, G.; Yeou, S.; Son, J.B.; Shin, Y.-K.; Lee, N.K. Cooperative inhibition of SNARE-mediated vesicle fusion by α-synuclein monomers and oligomers. *Sci. Rep.* **2021**, *11*, 10955. [CrossRef]
168. Alza, N.P.; Iglesias González, P.A.; Conde, M.A.; Uranga, R.M.; Salvador, G.A. Lipids at the Crossroad of α-Synuclein Function and Dysfunction: Biological and Pathological Implications. *Front. Cell. Neurosci.* **2019**, *13*, 175. [CrossRef]
169. Medeiros, A.T.; Soll, L.G.; Tessari, I.; Bubacco, L.; Morgan, J.R. α-Synuclein Dimers Impair Vesicle Fission during Clathrin-Mediated Synaptic Vesicle Recycling. *Front. Cell. Neurosci.* **2017**, *11*, 388. [CrossRef]
170. Fusco, G.; Chen, S.W.; Williamson, P.T.F.; Cascella, R.; Perni, M.; Jarvis, J.A.; Cecchi, C.; Vendruscolo, M.; Chiti, F.; Cremades, N.; et al. Structural basis of membrane disruption and cellular toxicity by α-synuclein oligomers. *Science* **2017**, *358*, 1440–1443. [CrossRef]
171. Mor, D.E.; Tsika, E.; Mazzulli, J.R.; Gould, N.S.; Kim, H.; Daniels, M.J.; Doshi, S.; Gupta, P.; Grossman, J.L.; Tan, V.X.; et al. Dopamine induces soluble α-synuclein oligomers and nigrostriatal degeneration. *Nat. Neurosci.* **2017**, *20*, 1560–1568. [CrossRef] [PubMed]
172. Kiechle, M.; Grozdanov, V.; Danzer, K.M. The Role of Lipids in the Initiation of α-Synuclein Misfolding. *Front. Cell. Dev. Biol.* **2020**, *8*, 562241. [CrossRef]
173. Tozzi, A.; Sciaccaluga, M.; Loffredo, V.; Megaro, A.; Ledonne, A.; Cardinale, A.; Federici, M.; Bellingacci, L.; Paciotti, S.; Ferrari, E.; et al. Dopamine-dependent early synaptic and motor dysfunctions induced by α-synuclein in the nigrostriatal circuit. *Brain* **2021**, 1–34. [CrossRef]
174. Matuskey, D.; Tinaz, S.; Wilcox, K.C.; Naganawa, M.; Toyonaga, T.; Dias, M.; Henry, S.; Pittman, B.; Ropchan, J.; Nabulsi, N.; et al. Synaptic Changes in Parkinson Disease Assessed with in vivo Imaging. *Ann. Neurol.* **2020**, *87*, 329–338. [CrossRef]
175. Soukup, S.; Vanhauwaert, R.; Verstreken, P. Parkinson's disease: Convergence on synaptic homeostasis. *EMBO J.* **2018**, *37*, 1–16. [CrossRef]
176. Candelise, N.; Schmitz, M.; Thüne, K.; Cramm, M.; Rabano, A.; Zafar, S.; Stoops, E.; Vanderstichele, H.; Villar-Pique, A.; Llorens, F.; et al. Effect of the micro-environment on α-synuclein conversion and implication in seeded conversion assays. *Transl. Neurodegener.* **2020**, *9*, 5. [CrossRef]
177. Perrin, R.J.; Woods, W.S.; Clayton, D.F.; George, J.M. Interaction of human alpha-Synuclein and Parkinson's disease variants with phospholipids. Structural analysis using site-directed mutagenesis. *J. Biol. Chem.* **2000**, *275*, 34393–34398. [CrossRef]
178. Jensen, P.H.; Nielsen, M.S.; Jakes, R.; Dotti, C.G.; Goedert, M. Binding of alpha-synuclein to brain vesicles is abolished by familial Parkinson's disease mutation. *J. Biol. Chem.* **1998**, *273*, 26292–26294. [CrossRef]
179. McLean, P.J.; Kawamata, H.; Ribich, S.; Hyman, B.T. Membrane association and protein conformation of alpha-synuclein in intact neurons. Effect of Parkinson's disease-linked mutations. *J. Biol. Chem.* **2000**, *275*, 8812–8816. [CrossRef]
180. Ruf, V.C.; Nübling, G.S.; Willikens, S.; Shi, S.; Schmidt, F.; Levin, J.; Bötzel, K.; Kamp, F.; Giese, A. Different Effects of α-Synuclein Mutants on Lipid Binding and Aggregation Detected by Single Molecule Fluorescence Spectroscopy and ThT Fluorescence-Based Measurements. *ACS Chem. Neurosci.* **2019**, *10*, 1649–1659. [CrossRef]

181. Fanning, S.; Selkoe, D.; Dettmer, U. Parkinson's disease: Proteinopathy or lipidopathy? *NPJ Park. Dis.* **2020**, *6*, 3. [CrossRef]
182. Killinger, B.A.; Melki, R.; Brundin, P.; Kordower, J.H. Endogenous alpha-synuclein monomers, oligomers and resulting pathology: Let's talk about the lipids in the room. *NPJ Park. Dis.* **2019**, *5*, 23. [CrossRef]
183. Mahul-Mellier, A.-L.; Burtscher, J.; Maharjan, N.; Weerens, L.; Croisier, M.; Kuttler, F.; Leleu, M.; Knott, G.W.; Lashuel, H.A. The process of Lewy body formation, rather than simply α-synuclein fibrillization, is one of the major drivers of neurodegeneration. *Proc. Natl. Acad. Sci. USA* **2020**, *117*, 4971–4982. [CrossRef]
184. Burré, J.; Sharma, M.; Südhof, T.C. α-Synuclein assembles into higher-order multimers upon membrane binding to promote SNARE complex formation. *Proc. Natl. Acad. Sci. USA* **2014**, *111*, E4274–E4283. [CrossRef]
185. Anwar, S.; Peters, O.; Millership, S.; Ninkina, N.; Doig, N.; Connor-Robson, N.; Threlfell, S.; Kooner, G.; Deacon, R.M.; Bannerman, D.M.; et al. Functional alterations to the nigrostriatal system in mice lacking all three members of the synuclein family. *J. Neurosci.* **2011**, *31*, 7264–7274. [CrossRef]
186. Reynolds, N.P.; Soragni, A.; Rabe, M.; Verdes, D.; Liverani, E.; Handschin, S.; Riek, R.; Seeger, S. Mechanism of Membrane Interaction and Disruption by α-Synuclein. *J. Am. Chem. Soc.* **2011**, *133*, 19366–19375. [CrossRef]
187. Suzuki, M.; Sango, K.; Wada, K.; Nagai, Y. Pathological role of lipid interaction with α-synuclein in Parkinson's disease. *Neurochem. Int.* **2018**, *119*, 97–106. [CrossRef]
188. Lee, J.Y.; Jin, H.K.; Bae, J.S. Sphingolipids in neuroinflammation: A potential target for diagnosis and therapy. *BMB Rep.* **2020**, *53*, 28–34. [CrossRef] [PubMed]
189. Hirsch, E.C.; Standaert, D.G. Ten Unsolved Questions About Neuroinflammation in Parkinson's Disease. *Mov. Disord.* **2021**, *36*, 16–24. [CrossRef] [PubMed]
190. Phan, J.-A.; Stokholm, K.; Zareba-Paslawska, J.; Jakobsen, S.; Vang, K.; Gjedde, A.; Landau, A.M.; Romero-Ramos, M. Early synaptic dysfunction induced by α-synuclein in a rat model of Parkinson's disease. *Sci. Rep.* **2017**, *7*, 6363. [CrossRef] [PubMed]
191. Takao, M.; Aoyama, M.; Ishikawa, K.; Sakiyama, Y.; Yomono, H.; Saito, Y.; Kurisaki, H.; Mihara, B.; Murayama, S. Spinocerebellar ataxia type 2 is associated with Parkinsonism and Lewy body pathology. *BMJ Case Rep.* **2011**, *2011*, bcr0120113685. [CrossRef]
192. Sen, N.-E.; Arsovic, A.; Meierhofer, D.; Brodesser, S.; Oberschmidt, C.; Canet-Pons, J.; Kaya, Z.-E.; Halbach, M.-V.; Gispert, S.; Sandhoff, K.; et al. In Human and Mouse Spino-Cerebellar Tissue, Ataxin-2 Expansion Affects Ceramide-Sphingomyelin Metabolism. *Int. J. Mol. Sci.* **2019**, *20*, 5854. [CrossRef]
193. Hogarth, P.; Gregory, A.; Kruer, M.C.; Sanford, L.; Wagoner, W.; Natowicz, M.R.; Egel, R.T.; Subramony, S.H.; Goldman, J.G.; Berry-Kravis, E.; et al. New NBIA subtype: Genetic, clinical, pathologic, and radiographic features of MPAN. *Neurology* **2013**, *80*, 268–275. [CrossRef]
194. Hartig, M.B.; Iuso, A.; Haack, T.; Kmiec, T.; Jurkiewicz, E.; Heim, K.; Roeber, S.; Tarabin, V.; Dusi, S.; Krajewska-Walasek, M.; et al. Absence of an orphan mitochondrial protein, c19orf12, causes a distinct clinical subtype of neurodegeneration with brain iron accumulation. *Am. J. Hum. Genet.* **2011**, *89*, 543–550. [CrossRef]
195. Chen, Y.P.; Song, W.; Huang, R.; Chen, K.; Zhao, B.; Li, J.; Yang, Y.; Shang, H.-F. GAK rs1564282 and DGKQ rs11248060 increase the risk for Parkinson's disease in a Chinese population. *J. Clin. Neurosci.* **2013**, *20*, 880–883. [CrossRef]
196. Nalls, M.A.; Pankratz, N.; Lill, C.M.; Do, C.B.; Hernandez, D.G.; Saad, M.; DeStefano, A.L.; Kara, E.; Bras, J.; Sharma, M.; et al. Large-scale meta-analysis of genome-wide association data identifies six new risk loci for Parkinson's disease. *Nat. Genet.* **2014**, *46*, 989–993. [CrossRef]
197. Bidinosti, M.; Shimshek, D.R.; Mollenhauer, B.; Marcellin, D.; Schweizer, T.; Lotz, G.P.; Schlossmacher, M.G.; Weiss, A. Novel one-step immunoassays to quantify α-synuclein: Applications for biomarker development and high-throughput screening. *J. Biol. Chem.* **2012**, *287*, 33691–33705. [CrossRef]
198. Chang, D.; Nalls, M.A.; Hallgrímsdóttir, I.B.; Hunkapiller, J.; van der Brug, M.; Cai, F.; Kerchner, G.A.; Ayalon, G.; Bingol, B.; Sheng, M.; et al. A meta-analysis of genome-wide association studies identifies 17 new Parkinson's disease risk loci. *Nat. Genet.* **2017**, *49*, 1511–1516. [CrossRef]
199. Li, G.; Cui, S.; Du, J.; Liu, J.; Zhang, P.; Fu, Y.; He, Y.; Zhou, H.; Ma, J.; Chen, S. Association of GALC, ZNF184, IL1R2 and ELOVL7 With Parkinson's Disease in Southern Chinese. *Front. Aging Neurosci.* **2018**, *10*, 402. [CrossRef]
200. Marshall, M.S.; Issa, Y.; Heller, G.; Nguyen, D.; Bongarzone, E.R. AAV-Mediated GALC Gene Therapy Rescues Alpha-Synucleinopathy in the Spinal Cord of a Leukodystrophic Lysosomal Storage Disease Mouse Model. *Front. Cell. Neurosci.* **2020**, *14*, 619712. [CrossRef]
201. Marshall, M.S.; Jakubauskas, B.; Bogue, W.; Stoskute, M.; Hauck, Z.; Rue, E.; Nichols, M.; DiAntonio, L.L.; van Breemen, R.B.; Kordower, J.H.; et al. Analysis of age-related changes in psychosine metabolism in the human brain. *PLoS ONE* **2018**, *13*, e0193438. [CrossRef]
202. Smith, B.R.; Santos, M.B.; Marshall, M.S.; Cantuti-Castelvetri, L.; Lopez-Rosas, A.; Li, G.; van Breemen, R.; Claycomb, K.I.; Gallea, J.I.; Celej, M.S.; et al. Neuronal inclusions of α-synuclein contribute to the pathogenesis of Krabbe disease. *J. Pathol.* **2014**, *232*, 509–521. [CrossRef]
203. Sidransky, E.; Nalls, M.A.; Aasly, J.O.; Aharon-Peretz, J.; Annesi, G.; Barbosa, E.R.; Bar-Shira, A.; Berg, D.; Bras, J.; Brice, A.; et al. Multicenter Analysis of Glucocerebrosidase Mutations in Parkinson's Disease. *N. Engl. J. Med.* **2009**, *361*, 1651–1661. [CrossRef]
204. Klemann, C.J.H.M.; Martens, G.J.M.; Sharma, M.; Martens, M.B.; Isacson, O.; Gasser, T.; Visser, J.E.; Poelmans, G. Integrated molecular landscape of Parkinson's disease. *NPJ Park. Dis.* **2017**, *3*, 14. [CrossRef]

205. Paisan-Ruiz, C.; Bhatia, K.P.; Li, A.; Hernandez, D.; Davis, M.; Wood, N.W.; Hardy, J.; Houlden, H.; Singleton, A.; Schneider, S.A. Characterization of PLA2G6 as a locus for dystonia-parkinsonism. *Ann. Neurol.* **2009**, *65*, 19–23. [CrossRef]
206. Kinghorn, K.J.; Castillo-Quan, J.I.; Bartolome, F.; Angelova, P.R.; Li, L.; Pope, S.; Cochemé, H.M.; Khan, S.; Asghari, S.; Bhatia, K.P.; et al. Loss of PLA2G6 leads to elevated mitochondrial lipid peroxidation and mitochondrial dysfunction. *Brain* **2015**, *138*, 1801–1816. [CrossRef]
207. Sumi-Akamaru, H.; Beck, G.; Shinzawa, K.; Kato, S.; Riku, Y.; Yoshida, M.; Fujimura, H.; Tsujimoto, Y.; Sakoda, S.; Mochizuki, H. High expression of α-synuclein in damaged mitochondria with PLA2G6 dysfunction. *Acta Neuropathol. Commun.* **2016**, *4*, 27. [CrossRef]
208. Do, C.B.; Tung, J.Y.; Dorfman, E.; Kiefer, A.K.; Drabant, E.M.; Francke, U.; Mountain, J.L.; Goldman, S.M.; Tanner, C.M.; Langston, J.W.; et al. Web-Based Genome-Wide Association Study Identifies Two Novel Loci and a Substantial Genetic Component for Parkinson's Disease. *PLoS Genet.* **2011**, *7*, e1002141. [CrossRef]
209. Gan-Or, Z.; Dion, P.A.; Rouleau, G.A. Genetic perspective on the role of the autophagy-lysosome pathway in Parkinson disease. *Autophagy* **2015**, *11*, 1443–1457. [CrossRef]
210. Lesage, S.; Drouet, V.; Majounie, E.; Deramecourt, V.; Jacoupy, M.; Nicolas, A.; Cormier-Dequaire, F.; Hassoun, S.M.; Pujol, C.; Ciura, S.; et al. Loss of VPS13C Function in Autosomal-Recessive Parkinsonism Causes Mitochondrial Dysfunction and Increases PINK1/Parkin-Dependent Mitophagy. *Am. J. Hum. Genet.* **2016**, *98*, 500–513. [CrossRef]
211. Kumar, N.; Leonzino, M.; Hancock-Cerutti, W.; Horenkamp, F.A.; Li, P.; Lees, J.A.; Wheeler, H.; Reinisch, K.M.; De Camilli, P. VPS13A and VPS13C are lipid transport proteins differentially localized at ER contact sites. *J. Cell Biol.* **2018**, *217*, 3625–3639. [CrossRef]
212. Kellie, J.F.; Higgs, R.E.; Ryder, J.W.; Major, A.; Beach, T.G.; Adler, C.H.; Merchant, K.; Knierman, M.D. Quantitative Measurement of Intact Alpha-Synuclein Proteoforms from Post-Mortem Control and Parkinson's Disease Brain Tissue by Intact Protein Mass Spectrometry. *Sci. Rep.* **2015**, *4*, 5797. [CrossRef]
213. Chakroun, T.; Evsyukov, V.; Nykänen, N.-P.; Höllerhage, M.; Schmidt, A.; Kamp, F.; Ruf, V.C.; Wurst, W.; Rösler, T.W.; Höglinger, G.U. Alpha-synuclein fragments trigger distinct aggregation pathways. *Cell Death Dis.* **2020**, *11*, 84. [CrossRef]
214. Imberdis, T.; Negri, J.; Ramalingam, N.; Terry-Kantor, E.; Ho, G.P.H.; Fanning, S.; Stirtz, G.; Kim, T.-E.; Levy, O.A.; Young-Pearse, T.L.; et al. Cell models of lipid-rich α-synuclein aggregation validate known modifiers of α-synuclein biology and identify stearoyl-CoA desaturase. *Proc. Natl. Acad. Sci. USA* **2019**, *116*, 20760–20769. [CrossRef]
215. Iljina, M.; Tosatto, L.; Choi, M.L.; Sang, J.C.; Ye, Y.; Hughes, C.D.; Bryant, C.E.; Gandhi, S.; Klenerman, D. Arachidonic acid mediates the formation of abundant alpha-helical multimers of alpha-synuclein. *Sci. Rep.* **2016**, *6*, 33928. [CrossRef]
216. Itokazu, Y.; Fuchigami, T.; Morgan, J.C.; Yu, R.K. Intranasal infusion of GD3 and GM1 gangliosides down-regulates alpha-synuclein and controls tyrosine hydroxylase gene in a PD model mouse. *Mol. Ther.* **2021**, in press. [CrossRef]
217. Perni, M.; Galvagnion, C.; Maltsev, A.; Meisl, G.; Müller, M.B.D.; Challa, P.K.; Kirkegaard, J.B.; Flagmeier, P.; Cohen, S.I.A.; Cascella, R.; et al. A natural product inhibits the initiation of α-synuclein aggregation and suppresses its toxicity. *Proc. Natl. Acad. Sci. USA* **2017**, *114*, E1009–E1017. [CrossRef]
218. Walther, T.C.; Farese, R. V Lipid Droplets and Cellular Lipid Metabolism. *Annu. Rev. Biochem.* **2012**, *81*, 687–714. [CrossRef]
219. Cohen, S. Lipid Droplets as Organelles. *Int. Rev. Cell Mol. Biol.* **2018**, *337*, 83–110. [CrossRef]
220. Girard, V.; Jollivet, F.; Knittelfelder, O.; Arsac, J.; Chatelain, G. A non-canonical lipid droplet metabolism regulates the conversion of alpha-Synuclein to proteolytic resistant forms in neurons of a Drosophila model of Parkinson disease. *bioRxiv* **2021**. [CrossRef]
221. Schoeler, M.; Caesar, R. Dietary lipids, gut microbiota and lipid metabolism. *Rev. Endocr. Metab. Disord.* **2019**, *20*, 461–472. [CrossRef]
222. Gérard, P. The crosstalk between the gut microbiota and lipids. *OCL* **2020**, *27*, 70. [CrossRef]
223. Devos, D.; Moreau, C.; Devedjian, J.C.; Kluza, J.; Petrault, M.; Laloux, C.; Jonneaux, A.; Ryckewaert, G.; Garçon, G.; Rouaix, N.; et al. Targeting Chelatable Iron as a Therapeutic Modality in Parkinson's Disease. *Antioxid. Redox. Signal.* **2014**, *21*, 195–210. [CrossRef]
224. Mahoney-Sánchez, L.; Bouchaoui, H.; Ayton, S.; Devos, D.; Duce, J.A.; Devedjian, J.-C. Ferroptosis and its potential role in the physiopathology of Parkinson's Disease. *Prog. Neurobiol.* **2021**, *196*, 101890. [CrossRef] [PubMed]
225. Caló, L.; Hidari, E.; Wegrzynowicz, M.; Dalley, J.W.; Schneider, B.L.; Podgajna, M.; Anichtchik, O.; Carlson, E.; Klenerman, D.; Spillantini, M.G. CSPα reduces aggregates and rescues striatal dopamine release in α-synuclein transgenic mice. *Brain* **2021**, *144*, 1661–1669. [CrossRef]

 cells 

Review

Hallmarks and Molecular Tools for the Study of Mitophagy in Parkinson's Disease

Thomas Goiran †, Mohamed A. Eldeeb †, Cornelia E. Zorca † and Edward A. Fon *

McGill Parkinson Program, Neurodegenerative Diseases Group, Department of Neurology and Neurosurgery, Montreal Neurological Institute, McGill University, Montreal, QC H3A 2B4, Canada; thomas.goiran@mail.mcgill.ca (T.G.); mohamed.eldeeb@mcgill.ca (M.A.E.); cornelia.zorca@mcgill.ca (C.E.Z.)
* Correspondence: ted.fon@mcgill.ca
† These authors contributed equally to this work.

Abstract: The best-known hallmarks of Parkinson's disease (PD) are the motor deficits that result from the degeneration of dopaminergic neurons in the substantia nigra. Dopaminergic neurons are thought to be particularly susceptible to mitochondrial dysfunction. As such, for their survival, they rely on the elaborate quality control mechanisms that have evolved in mammalian cells to monitor mitochondrial function and eliminate dysfunctional mitochondria. Mitophagy is a specialized type of autophagy that mediates the selective removal of damaged mitochondria from cells, with the net effect of dampening the toxicity arising from these dysfunctional organelles. Despite an increasing understanding of the molecular mechanisms that regulate the removal of damaged mitochondria, the detailed molecular link to PD pathophysiology is still not entirely clear. Herein, we review the fundamental molecular pathways involved in PINK1/Parkin-mediated and receptor-mediated mitophagy, the evidence for the dysfunction of these pathways in PD, and recently-developed state-of-the art assays for measuring mitophagy in vitro and in vivo.

Keywords: protein quality control; Parkinson's disease; mitochondrial quality control; ubiquitin; alpha-syn; mitophagy; PINK1; Parkin; mito-Keima; mito-QC; mito-SRAI

Citation: Goiran, T.; Eldeeb, M.A.; Zorca, C.E.; Fon, E.A. Hallmarks and Molecular Tools for the Study of Mitophagy in Parkinson's Disease. *Cells* **2022**, *11*, 2097. https://doi.org/10.3390/cells11132097

Academic Editor: Alexander E. Kalyuzhny

Received: 13 June 2022
Accepted: 29 June 2022
Published: 2 July 2022

1. Introduction

Mitochondria are essential organelles that possess their own genome and provide energy in the form of ATP for a variety of cellular processes [1–4]. For energy production, mitochondria rely on the process of oxidative phosphorylation (OXPHOS). The components of the OXPHOS machinery are encoded both in the nuclear and mitochondrial DNA. Dysfunction of OXPHOS components, especially of complex I, have been implicated with Parkinson's disease (PD), among other neurodegenerative diseases [5]. Early evidence of this came from observations that two complex I inhibitors, MPTP and rotenone, cause death of PD-associated dopamine neurons in both humans and rodent models [6,7]. Dysfunctional or otherwise damaged mitochondria are cleared by a specialized form of macroautophagy, called mitochondrial autophagy or mitophagy. A subset of sporadic forms of PD are thought to be associated with impaired mitochondrial function, though whether complex I defects are a cause or consequence of factors such as oxidative stress, is currently unclear. Likewise, familial forms of PD have been traced to mutations in genes encoding proteins associated with mitochondrial function and mitophagy, such as PINK1 and Parkin [8,9]. Moreover, in addition to genetic factors, environmental factors that affect mitochondria are also thought to play key roles in PD pathogenesis [10–12].

Cells possess several non-redundant mitophagy pathways; each can be triggered in response to different stimuli and each can elicit mitophagy through the activation of distinct signaling cascades (Figure 1) [13]. For instance, PINK1/Parkin dependent mitophagy is the main modulator of turnover of depolarized mitochondria. Additionally, several mitochondrial proteins, such as BNIP3, NIX, PHB2, and FUNDC1, have been shown to function as mitophagy receptors. These receptors are localized at the outer mitochondrial membrane (OMM) and interact directly with the autophagosomal membrane protein light chain 3 (LC3) to stimulate mitophagy. In addition, lipid-mediated mitophagy and ubiquitin-mediated mitophagy have also been reported [14–16]. Collectively, these pathways are deregulated in many human diseases, including neurodegenerative disorders, aging, and cancer [8,17–20]. In this review, we provide an overview of the key pathways involved in the regulation of mitophagy and their association with PD. We also discuss the molecular toolbox currently available to study this process in vitro and in vivo.

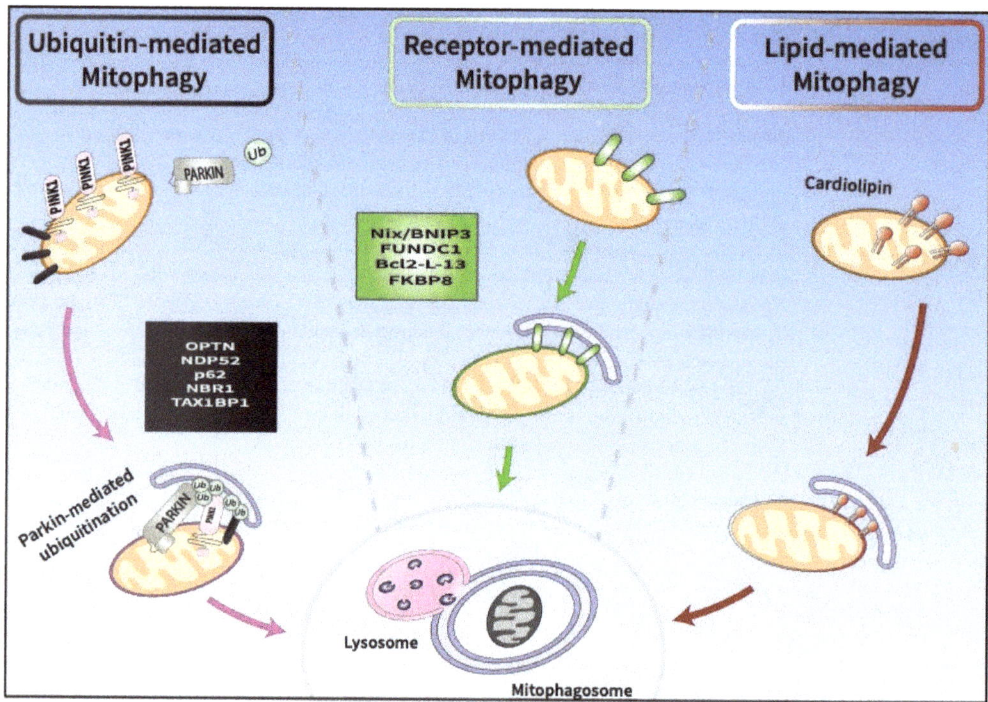

Figure 1. Three pathways at the crossroads of mitophagy. A. Ubiquitin-mediated mitophagy involves the recruitment of PINK1 and Parkin to the OMM, which promote the sequestration of damaged mitochondria into phagophores called mitophagosomes. Mitophagosomes subsequently fuse with lysosomes, where cargo is degraded. B. Receptor-mediated mitophagy depends on the direct binding of unique receptors, such as NIX/BNIP3L or FUNCD1, to LC3 on autophagosomes, which target damaged mitochondria for degradation. C. In lipid-mediated mitophagy, cardiolipin is externalized from the IMM to the OMM, where it binds to LC3 on mitophagosomes.

2. Molecular Pathways of Mitophagy

2.1. PINK1/Parkin-Mediated Ubiquitin-Driven Mitophagy

Ubiquitin (Ub) is a highly conserved small protein of 76 amino acids that is present in all eukaryotic cells. Ub plays a crucial role in many cellular processes, including protein degradation and immune system signaling. Ubiquitin-dependent protein degradation involves an enzymatic cascade resulting in the covalent conjugation of ubiquitin to the target protein substrate. This multi-step biochemical cascade leads either to the targeted degradation or to the altered localization of the substrate. The ubiquitination process is carried out by three enzymes: E1 (the ubiquitin activating enzyme), E2 (the ubiquitin conjugating enzyme, or carrier enzyme), and E3 (the ubiquitin protein-ligase). E3 Ub ligases are the principal factors that determine substrate specificity and are essential players in the ubiquitin pathway [21]. Parkin (encoded by the *PRKN* or *PARK2* gene) is an E3 Ub ligase, discovered in 1998, implicated in the pathogenesis of autosomal recessive PD (ARPD) [22–24]. Parkin contains five domains: an N-terminal Ub-like domain (UBL), a RING1 domain, an IBR domain, a RING2 domain, and a RING0 domain that is unique to Parkin [25–27]. Another ARPD-associated gene, PINK1 (PTEN-induced putative kinase 1), which is encoded by the *PARK6* gene, was discovered in 2001 [8,28] and encodes a mitochondrial serine/threonine kinase that regulates Parkin activity via phosphorylation. The PINK1 protein contains different domains, including an N-terminal mitochondrial targeting sequence (MTS), a transmembrane domain (TMD) followed by a serine/threonine kinase domain, and a regulatory domain at the C-terminus [29]. Under steady-state conditions, Parkin is located in the cytosol and is in an autoinhibited state. Concurrently, PINK1 is maintained at a low-level, owing to mitochondrial import, protease cleavage, and proteasomal degradation. Indeed, PINK1 is imported by the translocase of the outer membrane (TOM) complex into the inter membrane space (IMS) and the mitochondrial inner membrane (MIM), and then cleaved by the matrix processing peptidase (MPP) and the presenilin-associated rhomboid (PARL) in the N-terminal portion between Ala103 and Phe104. It is then retro-translocated to the cytosol, where the newly generated N-terminus, now consisting of the destabilizing amino acid Phe104, is constitutively recognized by N-end rule E3 ubiquitin ligases (UBR1, UBR2, and UBR4), leading to degradation of PINK1 by the proteasome [30–34]. However, reduction in the mitochondrial membrane potential results in the failure of PINK1 import and its accumulation on the outer mitochondrial membrane (OMM). This in turn leads to PINK1 dimerization and autophosphorylation at Ser228 and Ser402, resulting in its activation [35–38]. Thus, PINK1 accumulation on the OMM functions as a mitochondrial damage sensor that, once activated, triggers mitophagy and mediates downstream phosphorylation events, including the phosphorylation of ubiquitin at Ser65 and the phosphorylation of Ser65 in the UBL domain of Parkin. Phospho-ubiquitin (pSer65-Ub), is conjugated to proteins on the outer mitochondrial membrane; it then serves as a receptor for Parkin recruitment from the cytosol to mitochondria [36,39–42], and contributes to fully activating Parkin by inducing conformational changes in the Parkin core and releasing the UBL domain [43–49]. The precise mechanism by which Ubl phosphorylation activates Parkin is complex in nature, resulting in a number of Parkin activation models [43–49]. Subsequently, activated Parkin conjugates additional Ub moieties onto OMM proteins, marking the mitochondria for degradation by the autophagic machinery [18], thereby triggering mitophagy. Upon activation, Parkin polyubiquitinates several proteins on the OMM, including MFN1/2, TOM20/40/70, and VDAC 1 [50,51]. MFN1/2 (mitofusin), two GTPases required for mitochondrial fusion, were among the first and most crucial targets of Parkin-mediated ubiquitination. Mitochondria become progressively fragmented as a result of proteasomal degradation of MFN, resulting in their separation from one another. This phase appears to be crucial in distinguishing damaged mitochondrial fragments from the healthy reticulum that remains [49,50]. The ubiquitination of OMM proteins also facilitates the recruitment of receptor proteins that are part of the downstream autophagic degradation machinery (mitophagy). On the one hand, receptor proteins, such as p62, interact directly with polyubiquitin chains, and on the other hand, with LC3s or GABARAPs [51]. Initially, p62 was identified as the main adapter for PINK1/Parkin-mediated mitophagy [50]. Additional comprehensive studies identified five receptors: TAX1BP1, NDP52, NBR1, p62 and OPTN (Figure 2). Among these, NDP52 and OPTN were found to be the most important receptors for PINK1/Parkin-dependent mitophagy [52]. The recruitment of autophagy receptors, such as NDP52 and OPTN, to damaged mitochondria is a TANK-binding kinase 1- (TBK1) dependent process [52–54]. TBK1 is a serine/threonine kinase that enhances the binding ability of autophagy receptors to various Ub chains through their phosphorylation [52–54]. In the presence of PINK1 and Parkin, TBK1 activation also requires OPTN binding to Ub chains [53,54]. In the current mitophagy model, OPTN and NDP52 recruit phagophores to mitochondria by directly binding to LC3 through their LC3-interacting region (LIR), after binding to polyubiquitin chains [55,56]. A previous study has highlighted the role of NDP52 in the recruitment of the ULK1 complex to damaged mitochondria [57]. NDP52 directly interacts with FIP200 in a TBK1-dependent manner to recruit the ULK1 complex, leading to autophagosome biogenesis on damaged mitochondria and to the recruitment of the autophagy machinery. Therefore, receptor proteins ensure the removal of mitochondria by autophagosomes downstream of PINK1/Parkin signaling.

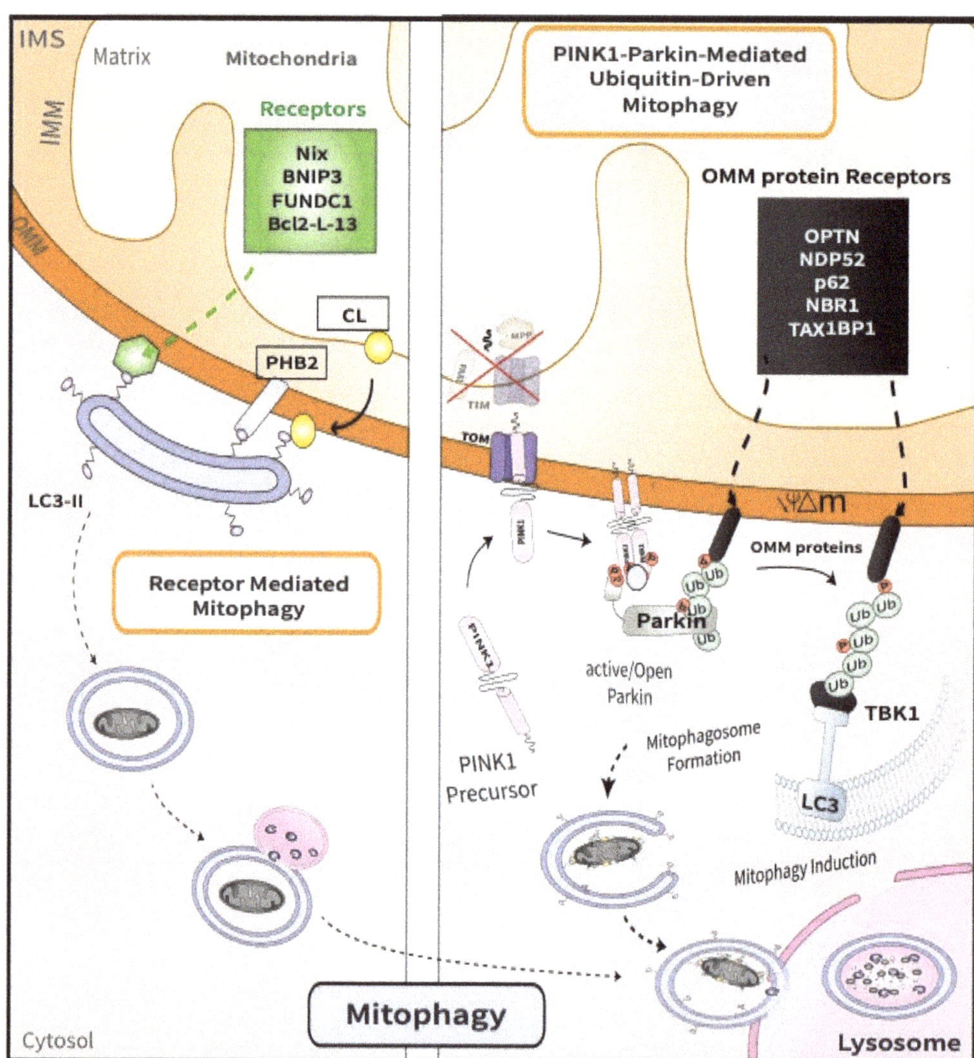

Figure 2. Comparison between PINK1/Parkin-mediated mitophagy and receptor-mediated mitophagy. The latter (**left**) involves the direct binding of mitophagy receptors to LC3 on the autophagosomes, which then deliver the engulfed damaged mitochondria to the lysosome. By contrast, the former (**right**) is a multi-step process that ensues following the loss of mitochondrial membrane potential. First, PINK1 is stabilized on the OMM of damaged mitochondria. Following dimerization, PINK1 recruits and phosphorylates Parkin, thereby initiating mitophagy.

2.2. Receptor Mediated Mitophagy

Several mitophagy receptors, such as ATG32 in yeast [58], as well as BNIP3 (BCL2 and adenovirus E1B 19-kDa-interacting protein 3) [59], NIX (also known as BNIP3L) [60], and FUNDC1 in mammalian cells, have been identified. One major characteristic of mitophagy receptors is that they contain LIR motifs that interact with LC3, thereby enhancing mitochondrial sequestration into phagophores [61–64]. The mechanism of BNIP3- and NIX-mediated mitophagy is distinct from that of the PINK1/Parkin pathway in that these proteins act as direct adaptors targeting mitochondria to the autophagosome. BNIP3 (a member of the pro-death BCL2 family of proteins) [65] and NIX (a homolog of BNIP3 with ~56% sequence similarity) [66] have a BH3 domain and a C-terminal transmembrane domain (TMD), which is crucial for their proapoptotic functions and mitochondrial localization [67,68]. Furthermore, BNIP3 and NIX have similar N-terminal LIR domains exposed to the cytosol that facilitates LC3s (microtubule-associated protein 1A/1B light chain) interactions for both receptors, or to GABARAP (gamma aminobutyric acid receptor-associated protein) for NIX, leading to the recruitment of autophagosomes and to the induction of mitophagy [61,69,70]. In these stress response pathways, the expression of BNIP3 is transcriptionally regulated by HIF−1, PPARγ, Rb/E2F, FoxO3, activated Ras, and p53, whereas that of NIX is regulated by HIF−1 and p53 [71–73]. Although BNIP3 and NIX are predominantly under transcriptional control, they are post-translationally modified for their mitophagic activity. Notably, it has been shown that serine phosphorylation at positions 17 and 24 adjacent to the LIR of BNIP3 and at positions 34 and 35 in the LIR domain of NIX enhances the interaction of these receptors with LC3, augmenting mitophagy [74].

Numerous lines of research suggest a possible crosstalk between the BNIP3/NIX receptor-mediated pathway and the PINK1/Parkin-mediated axis [75,76]. Specifically, NIX was implicated in PINK1/Parkin-mediated mitophagy as a ubiquitination substrate of Parkin that recruits NBR1 to the mitochondria [77]. Additionally, BNIP3-induced mitophagy is reduced in Parkin-deficient cells [78], and BNIP3 can stabilize PINK1 on the OMM, inhibiting its proteolytic degradation [79]. These results indicate that these pathways could cooperate with each other and may be partially redundant under particular cellular stress conditions to ensure effective mitophagy.

Another receptor-mediated mitophagy pathway hinges on the FUN14 domain containing 1 (FUNDC1). FUNDC1, an integral mitochondrial outer-membrane protein, is another important receptor for hypoxia-mediated mitophagy. FUNDC1 is composed of three transmembrane domains (TMDs) and an LIR domain in its N-terminus exposed to the cytosol, which interacts with LC3 for autophagosome recruitment [80]. Like other key regulators of mitophagy, the activity of FUNDC1 is also regulated by cycles of phosphorylation and dephosphorylation. The phosphorylation states of the three key residues, Ser13, Ser17 and Tyr18, in the outer membrane region of FUNDC1, play essential roles in fine-tuning the binding affinity for LC3 and controlling mitophagy [81,82]. Under steady-state conditions, the LIR motif of FUNDC1 is phosphorylated at Ser13 by CSNK2/CK2 kinase and at Tyr18 by SRC kinase, which leads to inhibition of its interaction with LC3 and prevents mitophagy. Conversely, hypoxia elicits dephosphorylation of FUNDC1, which can then bind to LC3 and elicit mitophagy [82]. Besides hypoxia, the array of cellular signals or states that can trigger receptor-mediated mitophagy remains to be fully elucidated [83–92].

3. Mitochondrial Defects in PD

3.1. Environmental Toxins as Risk Factors for PD

Among the mitochondrial defects associated with PD, reduced complex I activity has been found not only in the substantia nigra [83–85], but also in various other cells and tissues, including fibroblasts, lymphocytes, platelets, and in the skeletal muscle of sporadic PD patients [86–95]. However, mitochondrial complex I inhibition was shown to harm dopaminergic neurons more than other types of neurons [96]. The conditional ablation of an essential subunit of mitochondrial complex I, Ndufs2, in mouse dopaminergic neurons was recently shown to cause OXPHOS dysfunction and parkinsonian motor learning deficits that could be rescued by systemic levodopa administration [5]. Evidence of toxin-induced mitochondrial dysfunction has been recognized for over 30 years as a potential mechanism of dopaminergic neuronal loss associated with PD. Accidental exposure to MPTP (1-methyl−4-phenyl−1,2,3,6-tetrahydropyridine), a contaminant from the synthesis of MPPP (1-methyl−4-phenyl−4-propionoxy-piperidine), has been correlated with the rapid onset of parkinsonism [97]. Notably, MPTP itself is not toxic, but MPP+, its oxidized form, becomes toxic after being metabolized by mitochondrial monoamine oxidase B (MAO-B). MPP+ is selectively taken up into DA neurons through the dopamine transporter (DAT). Once internalized into neurons, MPP+ is rapidly concentrated in mitochondria [98–100], blocking electron transfer at complex I [101]. Such blockade results in the suppression of the complex I-mediated oxidation of nicotinamide adenine dinucleotide (NAD) and in OXPHOS dysfunction [102–104], thereby generating an abundance of free radicals (ROS), which has been proposed to contribute to DA neurodegeneration. Numerous studies have demonstrated that exposure to MPTP results in increased ROS levels, inhibition of mitochondrial respiration, DA neuron loss, and even cytoplasmic inclusions that share the characteristics of Lewy bodies (LB), the pathological hallmark of PD [105–111]. Interestingly, MPTP-treated mice that exhibited motor deficits and loss of TH expression in the substantia nigra, could be rescued by the co-administration of cell-permeable recombinant human Parkin [112]. Likewise, bypassing complex I and directly supplying the mitochondrial electron transport chain with complex II substrates enhanced OXPHOS and concomitantly reduced DA neurodegeneration in MPTP-treated mice [113]. Moreover, inhibition of complex I following MPTP treatment was shown to result in the degradation of the mitophagy receptor BNIP3L, in decreased protein ubiquitination, and in p62 inactivation [114–116], suggesting that impairments in both the ubiquitin-proteasome system and in the autophagic pathway can accompany mitophagy defects, in this context.

Other complex I inhibitors implicated in PD pathophysiology include rotenone and paraquat [117–119]. Rotenone is a lipophilic compound capable of crossing the blood-brain barrier, as well as cellular membranes. Like MPP+, rotenone inhibits complex I of the mitochondrial electron transport chain, resulting in increased ROS production, decreased ATP synthesis, and apoptotic cell death [117,120]. Increased ROS levels lead to mitochondrial dysfunction correlated with dopaminergic neuronal death [117,120]. In vivo proteomics studies have analyzed alterations in the striatum and substantia nigra caused by rotenone treatment [117,120]. Notably, the majority of altered proteins identified in these studies were involved in dopamine signaling, calcium signaling, apoptosis, and mitochondrial maintenance. Exposure to most of these PD environmental contaminants results in increased cellular ROS levels, inhibition of mitochondrial respiration, DA neuron loss, and LB-like inclusions [117,121,122].

3.2. Genetic Links to PD Risk

Over the past three decades, genetic studies have identified both dominantly and recessively inherited genes associated with familial forms of PD. Examples of the former include *SNCA* (*PARK1*) and *LRRK2* (*PARK8*), while examples of the latter include *PINK1*, and *PRKN*. Among these, *SNCA* and *LRRK2* have recently been associated with deficient mitochondrial function and homeostasis linked to PD pathophysiology. Specifically, mitochondrial α-synuclein accumulation was observed in a variety of neuronal and animal models, as well as in postmortem brain tissue of patients suffering from PD [95,123]. One way α-synuclein is thought to cause mitochondrial dysfunction is by binding to the translocase of the outer mitochondrial membrane 20 (TOMM20) and by inhibiting mitochondrial protein import [124]. Additionally, α-synuclein can directly inhibit complex I, complex IV, and ATP synthase, resulting in altered mitochondrial respiration and in mtDNA damage [125–129]. Recent evidence has implicated another familial PD gene, LRRK2, in the clearance of dysfunctional mitochondria. Hsieh et al. demonstrated that the pathogenic G2019S LRRK2 variant slowed the initiation of mitophagy in iPSC-derived neurons through a mechanism involving the delayed removal of a mitochondrial outer membrane protein, Miro1 [130]. Corroborating this, Singh et al. found that the hyperactive G2019S LRRK2 variant exhibited reduced mitophagy in dopaminergic neurons and microglia, which could be pharmacologically rescued by treatment with the GSK3357679A kinase inhibitor [131]. Lastly, increased levels of mitochondrial DNA (mtDNA) have been detected in the cerebrospinal fluid (CSF) of symptomatic G2019S LRRK2 carriers compared to asymptomatic carriers of this mutation [132]. While these findings identified mtDNA as a potential biomarker for LRRK2-associated PD, the link between circulating, cell-free mtDNA and mitochondrial dysfunction remains unclear [133].

Within the recessive category of genes, *PINK1* and *PRKN* have been shown to be directly involved in sensing and removing damaged mitochondria as described in the previous sections. Mutations in *PINK1* and *PRKN* have been associated with PD in different model systems. In Drosophila melanogaster, Caenorhabditis elegans, and Danio rerio (zebrafish), *PINK1* loss leads to anomalies in mitochondrial morphology and function, including decreased ATP production and increased ROS, as well as in neurodegeneration and locomotor deficits [134–136]. Germline *PINK1* knockout and *PRKN* knockout mice, on the other hand, show mitochondrial malfunction and increased sensitivity to oxidative stress, accompanied by minimal PD-like pathology [137–139]. However, upon aging, *PRKN* knockout mice were found to have both motor dysfunction and TH neuronal loss in the substantia nigra that correlated with the accumulation of damaged mitochondria within the dopaminergic neurons [140]. Likewise, *PRKN* knockout mice expressing a proofreading-defective DNA polymerase γ (POLG), which rapidly accumulate mtDNA mutations, exhibited mitochondrial abnormalities and dopaminergic neuronal loss [141,142]. Different rat models of PD were also found to recapitulate diverse pathological hallmarks, including mitochondrial dysfunction manifested by altered expression levels of complex I subunits in the striatum, deficits in complex I-driven respiration [143], and elevated levels of oxidative damage [144]. Lastly, midbrain dopaminergic neurons derived from induced pluripotent stem cells (iPSCs) harboring mutations in the *PINK1* or *PRKN* loci, exhibited both abnormal mitochondrial morphology and decreased survival upon mitochondrial stress induction with carbonyl cyanide m-chlorophenyl hydrazone (CCCP) [145]. In summary, these studies highlight the important roles of *PINK1* and *PRKN* in regulating mitochondrial function associated with PD pathogenesis.

In addition to familial studies, recent genome wide association studies (GWAS) have identified 90 genes as risk factors for PD [146–151]. While some of the loci implicated in monogenic familial PD have been shown to act directly in mitochondrial quality control and to play key roles in mitophagy, other GWAS genes are thought to exert indirect effects, most prominently affecting autophagy and lysosomal function [152–155]. For example, Inositol−1,4,5-triphosphate (IP3) kinase B (*ITPKB*) was shown to modulate mitochondrial ATP production through calcium released from the ER [156]. To complement the GWAS, single-cell transcriptomic analyses of different populations within the substantia nigra and cortex identified cell-specific gene networks associated with PD in post-mortem brain samples [157]. Prominent among these networks were groups of genes involved in mitochondrial organization, oxidative phosphorylation, and the electron transport chain [157]. Taken together, these unbiased studies demonstrate how gene alterations affect mitochondrial function in PD.

4. Mitophagy Assays

Diverse pathological conditions, including cancer, inflammatory, and neurodegenerative diseases, such as PD, have been associated with alterations in mitophagic capacity [158]. Consequently, the development of screens for compounds that modulate this fundamental cellular process holds tremendous translational potential through the discovery of novel drug targets [159]. Such screens rely on sensitive assays that measure mitophagy in both physiological and pathological conditions. Current well-established assays to monitor the selective removal of mitochondria measure different steps of the pathways described in the previous sections, and include: the quantification of endogenous or overexpressed Parkin or LC3 recruitment to the mitochondria and the quantification of the localization of mitochondria to the lysosomes compared to the cytosol [160,161]. Other methods of mitophagy detection assess mitochondrial alterations using fluorescent dyes, such as MitoTracker Deep Red or nonyl acridine orange (NAO), or by using transmission electron microscopy [162–164]. In addition, fluorescent reporters have been developed to measure the final steps of mitophagy, namely, the fusion of mitophagosomes with lysosomes, which we discuss in detail below (Figure 3).

4.1. mito-Keima

mito-Keima (mtKeima) is a pH-sensitive fluorescent biosensor that has been extensively used in a variety of cell lines, as well as in diverse model systems, including *Mus musculus* and *Drosophila melanogaster*, to measure mitophagy [52,165–168]. This reporter consists of a mutated version of the Keima protein found in stony corals, which has a pH-dependent excitation profile and a pH-insensitive emission peak at 620 nm [167,169,170]. Specifically, within a pH range of 6 to 8, which includes slightly alkali organelles, such as the mitochondria, the excitation maximum of Keima is at 440 nm [167,169,170]. By contrast, at acidic pH, which is one of the hallmarks of lysosomes, the excitation maximum of Keima shifts to 586 nm [167,169,170]. The mtKeima reporter is localized to the mitochondrial matrix by the cytochrome c oxidase subunit 8A (COX8A) targeting signal peptide sequence, artificially appended to the N-terminus of this fluorescent biosensor [167,169,170]. Consequently, mtKeima reporters found on healthy mitochondria exhibit an excitation/emission profile of 440 nm/620 nm, while those found on damaged mitochondria within autophagosomes that have fused with lysosomes have an excitation/emission profile of 586 nm/620 nm [167,169,170]. To assess the degree of mitochondrial clearance under steady-state versus pathological conditions, most often induced by protonophores such as CCCP or antimycin/oligomycin (OA), a 'mitophagy index' is calculated as the ratio of fluorescence intensity emitted from the two excitation peaks: 586 nm divided by 440 nm [167,169,170]. A high mitophagy index value indicates predominantly lysosomal localization of the biosensor, where it has been shown to remain remarkably insensitive to degradation by resident proteases [167,169,170]. Mitophagy index values have been obtained from different readouts, including single-cell fluorescence microscopy or flow cytometry amenable for analyzing large and diverse cell populations [167,169–171]. One drawback of mtKeima use is the intrinsic incompatibility of this method with fixation [159]. However, to date mtKeima has been extensively used as a robust reporter of in vivo mitophagy in different mammalian cell lines, including in induced pluripotent stem cell-derived dopaminergic neurons, and even in mice harboring a single-copy genomic integration of this reporter [52,166,167]. Notably, the latter demonstrated a remarkable degree of variability in the level of basal mitophagy among different cell types [167]. A small-scale chemical screen for modulators of mitophagy conducted with neural stem cells (NSCs) isolated from mtKeima transgenic mice identified actinonin as an inducer of this process [167]. This study demonstrated that the mtKeima mice are amenable, not only to a wide range of phenotypic studies, but also to pharmacological screens with cells isolated from diverse tissues [167].

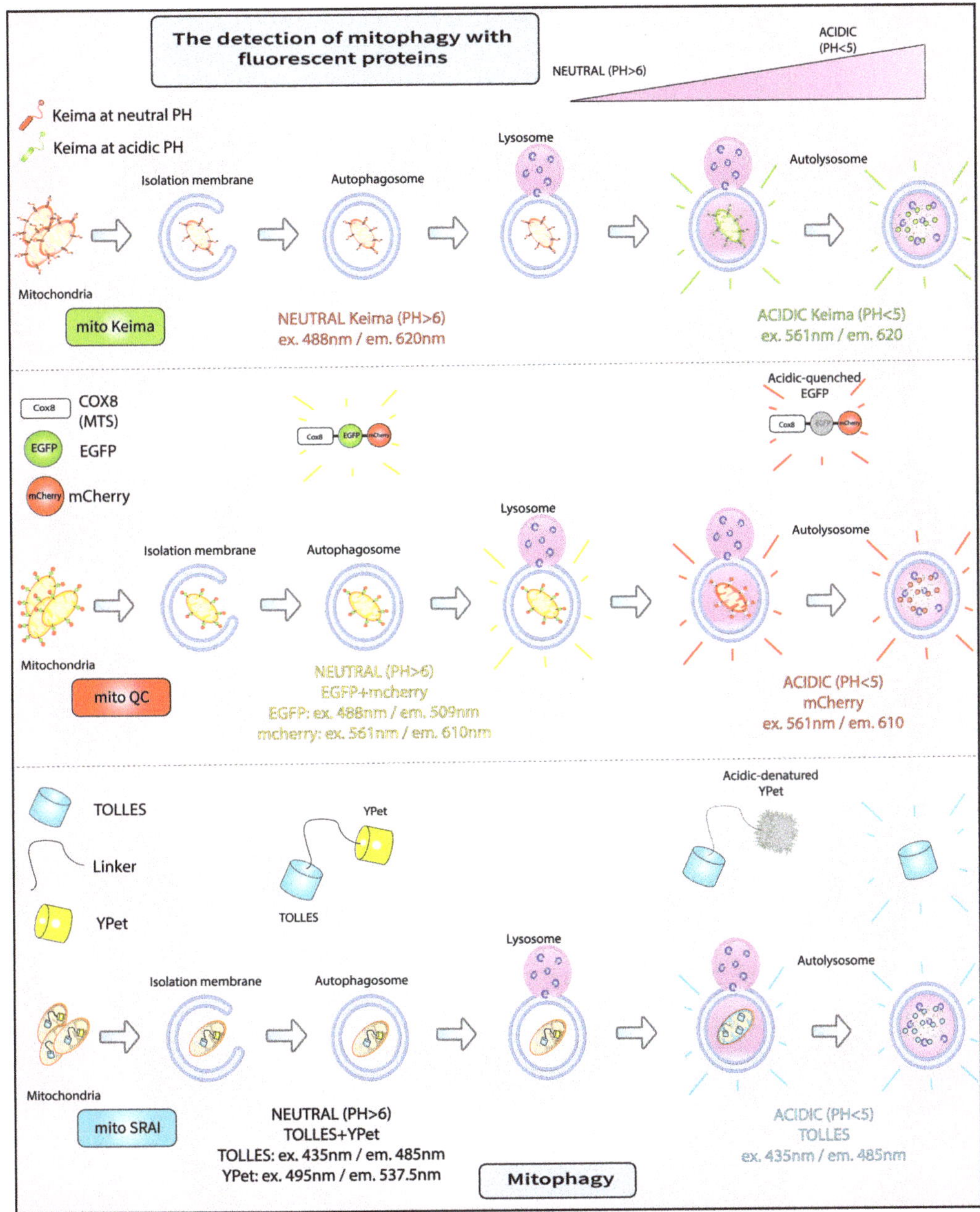

Figure 3. Fluorescent assays for mitophagy. A. mito-Keima is a pH-sensitive fluorescent biosensor, which fluoresces green at neutral pH in the cytosol, and red upon entry into acidic autolysosomes. B. mito-QC comprises two mitochondrially-targeted tandem fluorescent proteins, EGFP and mCherry. Both EGFP and mCherry fluoresce in the cytosol. However, in the lysosome, the fluorescence of mCherry is retained, while that of GFP is lost. C. mito-SRAI consists of two mitochondrially-targeted tandem fluorescent proteins, TOLLES and YPet. Unlike YPet, which is pH sensitive, TOLLES evades both acid-denaturation and proteolysis inside the lysosomal lumen and retains fluorescence.

4.2. mito-QC

An alternative probe to mtKeima is mito-QC. The design of this probe relies on two tandem fluorophores, mCherry and GFP, directed to the outer mitochondrial membrane by the C-terminal FIS1 transmembrane domain [172]. In the cytosol, both components of mito-QC fluoresce red and green, respectively, but when exposed to the low pH of the lysosomal compartment, the fluorescence of mCherry is maintained, while that of GFP is irreversibly lost [172]. In this case, a 'mitophagy index' is calculated as the count of exclusively red intracellular puncta, irrespective of size or intensity, since these puncta are thought to be resistant to lysosomal proteolysis [172]. One caveat of this approach is that mCherry and GFP have different maturation kinetics, with the former being slower than the latter, and different sensitivities to proteasomal degradation [173,174]. Unlike mtKeima, mito-QC can withstand fixation, a useful feature for colocalization with various cellular markers and analysis by fluorescence microscopy [172]. In addition to single-cell fluorescence microscopy, mito-QC has been analyzed by flow cytometry [175]. While numerous studies have used mito-QC to assay mitophagy in cell lines as well as in mice, only recently were these two reporters compared side-by-side [172,175–177]. The conclusions Liu et al. drew concerning the differential sensitivity level of mtKeima versus mito-QC as readouts for PINK1-Parkin-dependent mitophagy have raised debate in light of the intrinsic differences of these reporters [175,178]. Notably, systematic mito-QC analyses of basal mitophagy in different tissues isolated from *PINK1* wild type and knockout mice did not reveal remarkable differences, suggesting that PINK1 is not required for this type of mitophagy in vivo [179]. Another controversial concept in the field of mitophagy, which has been analyzed with the mito-QC and mtKeima reporters, revolves around soma-localized and axonal mitochondria. In particular, whether mitochondria in these compartments are differentially susceptible to mitophagy is still incompletely understood [180–182]. A sophisticated study by Harbauer et al. has begun to address this issue, demonstrating that axonal mitochondria undergo local PINK1/Parkin-mediated mitophagy [183]. Local translation of PINK1 mRNA, tethered to axonal mitochondria via Synaptojanin 2, is thought to facilitate mitophagy in distal axons by providing a supply of this labile protein [183]. Whether local translation of PINK1 mRNA occurs on mitochondrial ribosomes (mitoribosomes) remains to be directly elucidated. Nevertheless, this process circumvents the need for protein transport over long distances from the cell body, and facilitates a rapid local response to organelle damage or to bioenergetic changes within axons [184].

4.3. mito-SRAI

The most recently developed reporter of mitophagy is mito-SRAI. mito-SRAI consists of two tandem fluorescent proteins, acid-fast CFP or Tolerance of Lysosomal Environments (TOLLES) connected by a linker to YPet, a YFP variant [173]. The unique feature of acid-fast CFP is that it evades both acid denaturation and proteolysis inside the lysosomal lumen [173]. Consequently, acid-fast CFP fluorescence is preserved, while YPet fluorescence is lost within the lysosomes. As with the other reporters described above, mito-SRAI was extensively engineered and targeted to mitochondria not only by an N-terminal COX8A targeting signal peptide sequence, but also by C-terminal CL1 and PEST degrons that ensure the removal of free cytosolic reporters [173]. A 'mitophagy index' is calculated as afCFP fluorescence divided by YPet fluorescence. A high index value resulting from YPet quenching is indicative of mitophagy [173]. Similar to mito-QC, and unlike mtKeima, mito-SRAI is not sensitive to fixation [173]. Moreover, unlike the other two reporters, mito-SRAI could be used to specifically measure the mitophagy of damaged mitochondria [173]. Applying the mito-SRAI reporter to a large-scale screen of 76,000 compounds, Katayama et al. found a hit, called T-271, that effectively induced mitophagy of damaged, but not normal mitochondria, in a Parkin-dependent manner [173]. Further work is necessary to determine the molecular mechanisms involved in the selection of damaged mitochondria as opposed to healthy ones.

5. Future Perspectives

The shared feature of the three mitophagy assays described above is that they all report on the terminal lysosomal node of the pathway, responsible for removing damaged mitochondria. The development of robust alternatives that monitor different steps of the mitophagy pathway, amenable to high throughput studies (HTSs) will not only help advance our understanding of the mechanisms that regulate this process, but may reveal new nodes of intervention for drug targeting [159]. In particular, assays that monitor the spatiotemporal recruitment of early mitophagy effectors to the OMM in response to physiological or non-physiological stimuli, or mitophagy assays that utilize endogenous, rather than artificial fluorescent reporter systems, would open new avenues for exploration. For instance, one can envision the development of a fluorescent assay that can trace the dimerization of the BNIP3L/NIX receptor, which has been shown to be required for the induction of mitophagy [185]. By comparison with forced monomeric BNIP3L/NIX mutants, BNIP3L/NIX wild type receptors capable of forming homodimers have been shown to recruit autophagosomes more efficiently, as measured by LC3A immunofluorescence [185]. In addition, mutational analyses of key residues involved in either BNIP3L/NIX dimerization or in BNIP3L/NIX phosphorylation were demonstrated to affect mitochondrial clearance upon CCCP treatment [185]. Consequently, this dimerization event could be exploited as a mitophagy readout, potentially through a split-green fluorescent protein (GFP) system to monitor receptor homodimer formation in single cells [186,187]. This approach has been used to successfully tag members of the G-protein coupled receptor (GPCR) family of cell surface receptors [188]. Briefly, this method employs two independently non-fluorescent GFP fragments to tag a target of interest, which becomes fluorescent upon the complementation and reconstitution of a functional GFP protein [187,188]. Of note, a variation of split-GFP assays, called bi-genomic mitochondrial split-GFP, which is spatially confined to mitochondria, has recently been reported [189]. Besides mitophagy receptors, other molecules could be used as indicators of mitophagy towards assay development. For example, a recent study demonstrated that cardiolipin is externalized from the IMM to the OMM in primary cortical neurons and recruits LC3 to mitochondria, thereby initiating mitophagy [16]. This externalization process could also, in principle, be exploited as a mitophagy assay readout in a fluorescence resonance energy transfer (FRET)-based assay between a GFP-labeled probe containing the cardiolipin binding domain of the mitochondrially-localized stomatin-like protein 2 (SLP-2) and RFP-LC3 [190]. In conclusion, developing screens based on assays that rely on different steps of the mitochondrial clearance pathway holds tremendous promise for finding ways to enhance mitophagy under different pathological conditions.

6. Conclusions

Mounting evidence from genetic, cellular, and clinical studies over the past three decades points to the crucial role of mitochondrial dysfunction and mitophagy defects in PD. High-throughput assays, coupled with unbiased chemical or genetic screens for factors that can modulate mitophagy in susceptible dopaminergic neurons, are valuable tools for advancing PD therapies. Likewise, employing these tools to examine mitophagy in other cell types within the CNS, as well as the newly discovered process of trans-mitophagy, whereby neuron-derived damaged mitochondria are taken up and degraded by astrocytes, could offer additional points of therapeutic intervention [191,192].

Author Contributions: T.G., M.A.E., C.E.Z. and E.A.F. conceptualized the ideas and reviewed the literature; T.G., M.A.E. and C.E.Z. wrote the text and generated the figures. All authors have read and agreed to the published version of the manuscript.

Funding: This research was funded by a CIHR Foundation grant (FDN–154301) and by a Canada Research Chair 1315 (Tier 1) in Parkinson's disease.

Acknowledgments: E.A.F. was supported by a CIHR Foundation grant (FDN–154301) and by a Canada Research Chair 1315 (Tier 1) in Parkinson's disease. T.G. and M.A.E. were supported by Parkinson Canada. M.A.E. is a CIHR Banting Fellow.

Conflicts of Interest: The authors declare no competing interests.

References

1. Harbauer, A.B.; Zahedi, R.P.; Sickmann, A.; Pfanner, N.; Meisinger, C. The Protein Import Machinery of Mitochondria—A Regulatory Hub in Metabolism, Stress, and Disease. *Cell Metab.* **2014**, *19*, 357–372. [CrossRef] [PubMed]
2. Lill, R. Function and biogenesis of iron–sulphur proteins. *Nature* **2009**, *460*, 831–838. [CrossRef] [PubMed]
3. Schmidt, O.; Pfanner, N.; Meisinger, C. Mitochondrial protein import: From proteomics to functional mechanisms. *Nat. Rev. Mol. Cell Biol.* **2010**, *11*, 655–667. [CrossRef]
4. Hou, J.; Eldeeb, M.; Wang, X. Beyond Deubiquitylation: USP30-Mediated Regulation of Mitochondrial Homeostasis. *Mitochondrial DNA Dis.* **2017**, *1038*, 133–148. [CrossRef]
5. González-Rodríguez, P.; Zampese, E.; Stout, K.A.; Guzman, J.N.; Ilijic, E.; Yang, B.; Tkatch, T.; Stavarache, M.A.; Wokosin, D.L.; Gao, L.; et al. Disruption of mitochondrial complex I induces progressive parkinsonism. *Nature* **2021**, *599*, 650–656. [CrossRef]
6. Meredith, G.E.; Rademacher, D.J. MPTP Mouse Models of Parkinson's Disease: An Update. *J. Parkinson's Dis.* **2011**, *1*, 19–33. [CrossRef] [PubMed]
7. Wen, S.; Aki, T.; Unuma, K.; Uemura, K. Chemically Induced Models of Parkinson's Disease: History and Perspectives for the Involvement of Ferroptosis. *Front. Cell. Neurosci.* **2020**, *14*, 581191. [CrossRef]
8. Valente, E.M.; Abou-Sleiman, P.M.; Caputo, V.; Muqit, M.M.K.; Harvey, K.; Gispert, S.; Ali, Z.; Del Turco, D.; Bentivoglio, A.R.; Healy, D.G.; et al. Hereditary Early-Onset Parkinson's Disease Caused by Mutations in PINK1. *Science* **2004**, *304*, 1158–1160. [CrossRef]
9. Kitada, T.; Asakawa, S.; Hattori, N.; Matsumine, H.; Yamamura, Y.; Minoshima, S.; Yokochi, M.; Mizuno, Y.; Shimizu, N. Mutations in the parkin gene cause autosomal recessive juvenile parkinsonism. *Nature* **1998**, *392*, 605–608. [CrossRef]
10. Pezzoli, G.; Cereda, E. Exposure to pesticides or solvents and risk of Parkinson disease. *Neurology* **2013**, *80*, 2035–2041. [CrossRef]
11. Polito, L.; Greco, A.; Seripa, D. Genetic Profile, Environmental Exposure, and Their Interaction in Parkinson's Disease. *Park. Dis.* **2016**, *2016*, 6465793. [CrossRef] [PubMed]
12. Ryan, B.J.; Hoek, S.; Fon, E.A.; Wade-Martins, R. Mitochondrial dysfunction and mitophagy in Parkinson's: From familial to sporadic disease. *Trends Biochem. Sci.* **2015**, *40*, 200–210. [CrossRef] [PubMed]
13. Palikaras, K.; Daskalaki, I.; Markaki, M.; Tavernarakis, N. Mitophagy and age-related pathologies: Development of new therapeutics by targeting mitochondrial turnover. *Pharmacol. Ther.* **2017**, *178*, 157–174. [CrossRef] [PubMed]
14. Villa, E.; Marchetti, S.; Ricci, J.-E. No Parkin Zone: Mitophagy without Parkin. *Trends Cell Biol.* **2018**, *28*, 882–895. [CrossRef] [PubMed]
15. Strappazzon, F.; Nazio, F.; Corrado, M.; Cianfanelli, V.; Romagnoli, A.; Fimia, G.M.; Campello, S.; Nardacci, R.; Piacentini, M.; Campanella, M.; et al. AMBRA1 is able to induce mitophagy via LC3 binding, regardless of PARKIN and p62/SQSTM1. *Cell Death Differ.* **2015**, *22*, 517. [CrossRef]
16. Chu, T.C.; Ji, J.; Dagda, R.K.; Jiang, J.F.; Tyurina, Y.Y.; Kapralov, A.A.; Tyurin, V.A.; Yanamala, N.; Shrivastava, I.H.; Mohammadyani, D.; et al. Cardiolipin externalization to the outer mitochondrial membrane acts as an elimination signal for mitophagy in neuronal cells. *Nat. Cell Biol.* **2013**, *15*, 1197–1205. [CrossRef]
17. Pickles, S.; Vigié, P.; Youle, R.J. Mitophagy and Quality Control Mechanisms in Mitochondrial Maintenance. *Curr. Biol.* **2018**, *28*, R170–R185. [CrossRef]
18. Narendra, D.; Tanaka, A.; Suen, D.F.; Youle, R.J. Parkin is recruited selectively to impaired mitochondria and promotes their autophagy. *J. Cell Biol.* **2008**, *183*, 795–803. [CrossRef]
19. Chourasia, A.H.; Tracy, K.; Frankenberger, C.; Boland, M.L.; Sharifi, M.N.; Drake, L.E.; Sachleben, J.R.; Asara, J.M.; Locasale, J.W.; Karczmar, G.S.; et al. Mitophagy defects arising from BNip3 loss promote mammary tumor progression to metastasis. *EMBO Rep.* **2015**, *16*, 1145–1163. [CrossRef]
20. Springer, M.Z.; MacLeod, K.F. In Brief: Mitophagy: Mechanisms and role in human disease. *J. Pathol.* **2016**, *240*, 253–255. [CrossRef]
21. Scheffner, M.; Nuber, U.; Huibregtse, J.M. Protein ubiquitination involving an E1-E2-E3 enzyme ubiquitin thioester cascade. *Nature* **1995**, *373*, 81–83. [CrossRef] [PubMed]
22. Lücking, C.B.; Abbas, N.; Dürr, A.; Bonifati, V.; Bonnet, A.M.; De Broucker, T.; De Michele, G.; Wood, N.W.; Agid, Y.; Brice, A.; et al. Homozygous deletions in parkin gene in European and North African families with autosomal recessive juvenile parkinsonism. *Lancet* **1998**, *352*, 1355–1356. [CrossRef]
23. Abbas, N.; Lücking, C.B.; Ricard, S.; Dürr, A.; Bonifati, V.; De Michele, G.; Bouley, S.; Vaughan, J.R.; Gasser, T.; Marconi, R.; et al. A wide variety of mutations in the parkin gene are responsible for autosomal recessive parkinsonism in Europe. French Parkinson's disease genetics study group and the european consortium on genetic susceptibility in Parkinson's Disease. *Hum. Mol. Genet* **1999**, *8*, 567–574. [CrossRef] [PubMed]
24. Arkinson, C.; Walden, H. Parkin function in Parkinson's disease. *Science* **2018**, *360*, 267–268. [CrossRef] [PubMed]

25. Hristova, V.A.; Beasley, S.A.; Rylett, R.J.; Shaw, G.S. Identification of a Novel Zn2+-binding Domain in the Autosomal Recessive Juvenile Parkinson-related E3 Ligase Parkin. *J. Biol. Chem.* **2009**, *284*, 14978–14986. [CrossRef] [PubMed]
26. Trempe, J.-F.; Sauvé, V.; Grenier, K.; Seirafi, M.; Tang, M.Y.; Ménade, M.; Al-Abdul-Wahid, S.; Krett, J.; Wong, K.; Kozlov, G.; et al. Structure of Parkin Reveals Mechanisms for Ubiquitin Ligase Activation. *Science* **2013**, *340*, 1451–1455. [CrossRef] [PubMed]
27. Walden, H.; Muqit, M.M. Ubiquitin and Parkinson's disease through the looking glass of genetics. *Biochem. J.* **2017**, *474*, 1439–1451. [CrossRef]
28. Unoki, M.; Nakamura, Y. Growth-suppressive effects of BPOZ and EGR2, two genes involved in the PTEN signaling pathway. *Oncogene* **2001**, *20*, 4457–4465. [CrossRef]
29. Okatsu, K.; Kimura, M.; Oka, T.; Tanaka, K.; Matsuda, N. Unconventional PINK1 localization mechanism to the outer membrane of depolarized mitochondria drives Parkin recruitment. *J. Cell Sci.* **2015**, *128*, 964–978. [CrossRef]
30. Jin, S.M.; Lazarou, M.; Wang, C.; Kane, L.A.; Narendra, D.P.; Youle, R.J. Mitochondrial membrane potential regulates PINK1 import and proteolytic destabilization by PARL. *J. Cell Biol.* **2010**, *191*, 933–942. [CrossRef]
31. Deas, E.; Plun-Favreau, H.; Gandhi, S.; Desmond, H.; Kjaer, S.; Loh, S.H.; Renton, A.E.; Harvey, R.J.; Whitworth, A.J.; Martins, L.M.; et al. PINK1 cleavage at position A103 by the mitochondrial protease PARL. *Hum. Mol. Genet.* **2010**, *20*, 867–879. [CrossRef] [PubMed]
32. Lazarou, M.; Jin, S.M.; Kane, L.A.; Youle, R.J. Role of PINK1 Binding to the TOM Complex and Alternate Intracellular Membranes in Recruitment and Activation of the E3 Ligase Parkin. *Dev. Cell* **2012**, *22*, 320–333. [CrossRef] [PubMed]
33. Greene, A.W.; Grenier, K.; Aguileta, M.A.; Muise, S.; Farazifard, R.; Haque, M.E.; McBride, H.M.; Park, D.S.; Fon, E.A. Mitochondrial processing peptidase regulates PINK1 processing, import and Parkin recruitment. *EMBO Rep.* **2012**, *13*, 378–385. [CrossRef]
34. Yamano, K.; Youle, R.J. PINK1 is degraded through the N-end rule pathway. *Autophagy* **2013**, *9*, 1758–1769. [CrossRef] [PubMed]
35. Okatsu, K.; Oka, T.; Iguchi, M.; Imamura, K.; Kosako, H.; Tani, N.; Kimura, M.; Go, E.; Koyano, F.; Funayama, M.; et al. PINK1 autophosphorylation upon membrane potential dissipation is essential for Parkin recruitment to damaged mitochondria. *Nat. Commun.* **2012**, *3*, 1016. [CrossRef] [PubMed]
36. Okatsu, K.; Uno, M.; Koyano, F.; Go, E.; Kimura, M.; Oka, T.; Tanaka, K.; Matsuda, N. A Dimeric PINK1-containing Complex on Depolarized Mitochondria Stimulates Parkin Recruitment. *J. Biol. Chem.* **2013**, *288*, 36372–36384. [CrossRef]
37. Aerts, L.; Craessaerts, K.; De Strooper, B.; Morais, V. PINK1 Kinase Catalytic Activity Is Regulated by Phosphorylation on Serines 228 and 402. *J. Biol. Chem.* **2015**, *290*, 2798–2811. [CrossRef] [PubMed]
38. Rasool, S.; Soya, N.; Truong, L.; Croteau, N.; Lukacs, G.L.; Trempe, J. PINK 1 autophosphorylation is required for ubiquitin recognition. *EMBO Rep.* **2018**, *19*, e44981. [CrossRef]
39. Kane, L.A.; Lazarou, M.; Fogel, A.I.; Li, Y.; Yamano, K.; Sarraf, S.A.; Banerjee, S.; Youle, R.J. PINK1 phosphorylates ubiquitin to activate Parkin E3 ubiquitin ligase activity. *J. Cell Biol.* **2014**, *205*, 143–153. [CrossRef]
40. Kazlauskaite, A.; Kondapalli, C.; Gourlay, R.; Campbell, D.G.; Ritorto, M.S.; Hofmann, K.; Alessi, D.R.; Knebel, A.; Trost, M.; Muqit, M.M.K. Parkin is activated by PINK1-dependent phosphorylation of ubiquitin at Ser^{65}. *Biochem. J.* **2014**, *460*, 127–139. [CrossRef]
41. Koyano, F.; Okatsu, K.; Kosako, H.; Tamura, Y.; Go, E.; Kimura, M.; Kimura, Y.; Tsuchiya, H.; Yoshihara, H.; Hirokawa, T.; et al. Ubiquitin is phosphorylated by PINK1 to activate parkin. *Nature* **2014**, *510*, 162–166. [CrossRef] [PubMed]
42. Shiba-Fukushima, K.; Arano, T.; Matsumoto, G.; Inoshita, T.; Yoshida, S.; Ishihama, Y.; Ryu, K.-Y.; Nukina, N.; Hattori, N.; Imai, Y. Phosphorylation of Mitochondrial Polyubiquitin by PINK1 Promotes Parkin Mitochondrial Tethering. *PLoS Genet.* **2014**, *10*, e1004861. [CrossRef] [PubMed]
43. Ordureau, A.; Heo, J.-M.; Duda, D.M.; Paulo, J.A.; Olszewski, J.L.; Yanishevski, D.; Rinehart, J.; Schulman, B.A.; Harper, J.W. Defining roles of PARKIN and ubiquitin phosphorylation by PINK1 in mitochondrial quality control using a ubiquitin replacement strategy. *Proc. Natl. Acad. Sci. USA* **2015**, *112*, 6637–6642. [CrossRef] [PubMed]
44. Kazlauskaite, A.; Martinez-Torres, R.J.; Wilkie, S.; Kumar, A.; Peltier, J.; Gonzalez, A.; Johnson, C.; Zhang, J.; Hope, A.G.; Peggie, M.; et al. Binding to serine 65-phosphorylated ubiquitin primes Parkin for optimal PINK 1-dependent phosphorylation and activation. *EMBO Rep.* **2015**, *16*, 939–954. [CrossRef]
45. Kumar, A.; Aguirre, J.; Condos, T.E.C.; Martinez-Torres, R.J.; Chaugule, V.; Toth, R.; Sundaramoorthy, R.; Mercier, P.; Knebel, A.; Spratt, D.; et al. Disruption of the autoinhibited state primes the E3 ligase parkin for activation and catalysis. *EMBO J.* **2015**, *34*, 2506–2521. [CrossRef]
46. Sauvé, V.; Lilov, A.; Seirafi, M.; Vranas, M.; Rasool, S.; Kozlov, G.; Sprules, T.; Wang, J.; Trempe, J.; Gehring, K. A Ubl/ubiquitin switch in the activation of Parkin. *EMBO J.* **2015**, *34*, 2492–2505. [CrossRef]
47. Ordureau, A.; Sarraf, S.A.; Duda, D.M.; Heo, J.-M.; Jedrychowski, M.P.; Sviderskiy, V.O.; Olszewski, J.L.; Koerber, J.T.; Xie, T.; Beausoleil, S.A.; et al. Quantitative Proteomics Reveal a Feedforward Mechanism for Mitochondrial PARKIN Translocation and Ubiquitin Chain Synthesis. *Mol. Cell* **2014**, *56*, 360–375. [CrossRef]
48. Wauer, T.; Simicek, M.; Schubert, A.; Komander, D. Mechanism of phospho-ubiquitin-induced PARKIN activation. *Nature* **2015**, *524*, 370–374. [CrossRef]
49. Harper, J.W.; Ordureau, A.; Heo, J.-M. Building and decoding ubiquitin chains for mitophagy. *Nat. Rev. Mol. Cell Biol.* **2018**, *19*, 93–108. [CrossRef]

50. Geisler, S.; Holmström, K.; Skujat, D.; Fiesel, F.; Rothfuss, O.C.; Kahle, P.J.; Springer, W. PINK1/Parkin-mediated mitophagy is dependent on VDAC1 and p62/SQSTM1. *Nat. Cell Biol.* **2010**, *12*, 119–131. [CrossRef]
51. Sarraf, S.; Raman, M.; Guarani-Pereira, V.; Sowa, M.E.; Huttlin, E.; Gygi, S.P.; Harper, J.W. Landscape of the PARKIN-dependent ubiquitylome in response to mitochondrial depolarization. *Nature* **2013**, *496*, 372–376. [CrossRef] [PubMed]
52. McLelland, G.-L.; Goiran, T.; Yi, W.; Dorval, G.; Chen, C.X.; Lauinger, N.D.; Krahn, A.I.; Valimehr, S.; Rakovic, A.; Rouiller, I.; et al. Mfn2 ubiquitination by PINK1/parkin gates the p97-dependent release of ER from mitochondria to drive mitophagy. *eLife* **2018**, *7*, e32866. [CrossRef] [PubMed]
53. Stolz, A.; Ernst, A.; Dikic, I. Cargo recognition and trafficking in selective autophagy. *Nature* **2014**, *16*, 495–501. [CrossRef]
54. Lazarou, M.; Sliter, D.A.; Kane, L.A.; Sarraf, S.A.; Wang, C.; Burman, J.L.; Sideris, D.P.; Fogel, A.I.; Youle, R.J. The ubiquitin kinase PINK1 recruits autophagy receptors to induce mitophagy. *Nature* **2015**, *524*, 309–314. [CrossRef] [PubMed]
55. Heo, J.-M.; Ordureau, A.; Paulo, J.A.; Rinehart, J.; Harper, J.W. The PINK1-PARKIN Mitochondrial Ubiquitylation Pathway Drives a Program of OPTN/NDP52 Recruitment and TBK1 Activation to Promote Mitophagy. *Mol. Cell* **2015**, *60*, 7–20. [CrossRef]
56. Richter, B.; Sliter, D.A.; Herhaus, L.; Stolz, A.; Wang, C.; Beli, P.; Zaffagnini, G.; Wild, P.; Martens, S.; Wagner, S.A.; et al. Phosphorylation of OPTN by TBK1 enhances its binding to Ub chains and promotes selective autophagy of damaged mitochondria. *Proc. Natl. Acad. Sci. USA* **2016**, *113*, 4039–4044. [CrossRef]
57. Gatica, D.; Lahiri, V.; Klionsky, D.J. Cargo recognition and degradation by selective autophagy. *Nat. Cell Biol.* **2018**, *20*, 233–242. [CrossRef]
58. Palikaras, K.; Lionaki, E.; Tavernarakis, N. Mechanisms of mitophagy in cellular homeostasis, physiology and pathology. *Nat. Cell Biol.* **2018**, *20*, 1013–1022. [CrossRef]
59. Vargas, J.N.S.; Wang, C.; Bunker, E.; Hao, L.; Maric, D.; Schiavo, G.; Randow, F.; Youle, R.J. Spatiotemporal Control of ULK1 Activation by NDP52 and TBK1 during Selective Autophagy. *Mol. Cell* **2019**, *74*, 347–362.e6. [CrossRef]
60. Okamoto, K.; Kondo-Okamoto, N.; Ohsumi, Y. Mitochondria-Anchored Receptor Atg32 Mediates Degradation of Mitochondria via Selective Autophagy. *Dev. Cell* **2009**, *17*, 87–97. [CrossRef]
61. Hanna, R.A.; Quinsay, M.N.; Orogo, A.M.; Giang, K.; Rikka, S.; Gustafsson, Å.B. Microtubule-associated Protein 1 Light Chain 3 (LC3) Interacts with Bnip3 Protein to Selectively Remove Endoplasmic Reticulum and Mitochondria via Autophagy. *J. Biol. Chem.* **2012**, *287*, 19094–19104. [CrossRef] [PubMed]
62. Chen, Y.; Lewis, W.; Diwan, A.; Cheng, E.H.-Y.; Matkovich, S.J.; Dorn, G.W. Dual autonomous mitochondrial cell death pathways are activated by Nix/BNip3L and induce cardiomyopathy. *Proc. Natl. Acad. Sci. USA* **2010**, *107*, 9035–9042. [CrossRef] [PubMed]
63. Wei, H.; Liu, L.; Chen, Q. Selective removal of mitochondria via mitophagy: Distinct pathways for different mitochondrial stresses. *Biochim. Biophys. Acta (BBA)-Bioenerg.* **2015**, *1853*, 2784–2790. [CrossRef]
64. Bhujabal, Z.; Birgisdottir, A.B.; Sjottem, E.; Brenne, H.B.; Øvervatn, A.; Habisov, S.; Kirkin, V.; Lamark, T.; Johansen, T. FKBP8 recruits LC3A to mediate Parkin-independent mitophagy. *EMBO Rep.* **2017**, *18*, 947–961. [CrossRef] [PubMed]
65. Boyd, J.M.; Malstrom, S.; Subramanian, T.; Venkatesh, L.; Schaeper, U.; Elangovan, B.; D'Sa-Eipper, C.; Chinnadurai, G. Adenovirus E1B 19 kDa and Bcl-2 proteins interact with a common set of cellular proteins. *Cell* **1994**, *79*, 341–351. [CrossRef]
66. Matsushima, M.; Fujiwara, T.; Takahashi, E.I.; Minaguchi, T.; Eguchi, Y.; Tsujimoto, Y.; Suzumori, K.; Nakamura, Y. Isolation, mapping, and functional analysis of a novel human cDNA (BNIP3L) encoding a protein homologous to human NIP3. *Genes Chromosomes Cancer* **1998**, *21*, 230–235. [CrossRef]
67. Yasuda, M.; Theodorakis, P.; Subramanian, T.; Chinnadurai, G. Adenovirus E1B-19K/BCL-2 Interacting Protein BNIP3 Contains a BH3 Domain and a Mitochondrial Targeting Sequence. *J. Biol. Chem.* **1998**, *273*, 12415–12421. [CrossRef]
68. Imazu, T.; Shimizu, S.; Tagami, S.; Matsushima, M.; Nakamura, Y.; Miki, T.; Okuyama, A.; Tsujimoto, Y. Bcl-2/E1B 19 kDa-interacting protein 3-like protein (Bnip3L) interacts with bcl-2/Bcl-xL and induces apoptosis by altering mitochondrial membrane permeability. *Oncogene* **1999**, *18*, 4523–4529. [CrossRef]
69. Novak, I.; Kirkin, V.; McEwan, D.G.; Zhang, J.; Wild, P.; Rozenknop, A.; Rogov, V.; Löhr, F.; Popovic, D.; Occhipinti, A.; et al. Nix is a selective autophagy receptor for mitochondrial clearance. *EMBO Rep.* **2009**, *11*, 45–51. [CrossRef]
70. Birgisdottir, Å.B.; Lamark, T.; Johansen, T. The LIR motif-crucial for selective autophagy. *J. Cell Sci.* **2013**, *126*, 3237–3247. [CrossRef]
71. Sowter, H.M.; Ratcliffe, P.J.; Watson, P.; Greenberg, A.H.; Harris, A.L. HIF-1-dependent regulation of hypoxic induction of the cell death factors BNIP3 and NIX in human tumors. *Cancer Res.* **2001**, *61*, 6669–6673. [PubMed]
72. Mammucari, C.; Milan, G.; Romanello, V.; Masiero, E.; Rudolf, R.; Del Piccolo, P.; Burden, S.J.; Di Lisi, R.; Sandri, C.; Zhao, J.; et al. FoxO3 Controls Autophagy in Skeletal Muscle In Vivo. *Cell Metab.* **2007**, *6*, 458–471. [CrossRef] [PubMed]
73. Zhang, H.; Bosch-Marce, M.; Shimoda, L.A.; Tan, Y.S.; Baek, J.H.; Wesley, J.B.; Gonzalez, F.J.; Semenza, G.L. Mitochondrial Autophagy Is an HIF-1-dependent Adaptive Metabolic Response to Hypoxia. *J. Biol. Chem.* **2008**, *283*, 10892–10903. [CrossRef] [PubMed]
74. Rogov, V.V.; Suzuki, H.; Marinković, M.; Lang, V.; Kato, R.; Kawasaki, M.; Buljubasic, M.; Šprung, M.; Rogova, N.; Wakatsuki, S.; et al. Phosphorylation of the mitochondrial autophagy receptor Nix enhances its interaction with LC3 proteins. *Sci. Rep.* **2017**, *7*, 1131. [CrossRef]
75. Ding, W.-X.; Ni, H.-M.; Li, M.; Liao, Y.; Chen, X.; Stolz, D.B.; Dorn, G.W., 2nd; Yin, X.-M. Nix Is Critical to Two Distinct Phases of Mitophagy, Reactive Oxygen Species-mediated Autophagy Induction and Parkin-Ubiquitin-p62-mediated Mitochondrial Priming. *J. Biol. Chem.* **2010**, *285*, 27879–27890. [CrossRef]

76. Choi, G.E.; Lee, H.J.; Chae, C.W.; Cho, J.H.; Jung, Y.H.; Kim, J.S.; Kim, S.Y.; Lim, J.R.; Han, H.J. BNIP3L/NIX-mediated mitophagy protects against glucocorticoid-induced synapse defects. *Nat. Commun.* **2021**, *12*, 487. [CrossRef]
77. Gao, F.; Chen, D.; Si, J.; Hu, Q.; Qin, Z.; Fang, M.; Wang, G. The mitochondrial protein BNIP3L is the substrate of PARK2 and mediates mitophagy in PINK1/PARK2 pathway. *Hum. Mol. Genet.* **2015**, *24*, 2528–2538. [CrossRef]
78. Lee, Y.; Lee, H.-Y.; Hanna, R.A.; Gustafsson, Å.B. Mitochondrial autophagy by Bnip3 involves Drp1-mediated mitochondrial fission and recruitment of Parkin in cardiac myocytes. *Am. J. Physiol.-Heart Circ. Physiol.* **2011**, *301*, H1924–H1931. [CrossRef]
79. Zhang, T.; Xue, L.; Li, L.; Tang, C.; Wan, Z.; Wang, R.; Tan, J.; Tan, Y.; Han, H.; Tian, R.; et al. BNIP3 Protein Suppresses PINK1 Kinase Proteolytic Cleavage to Promote Mitophagy. *J. Biol. Chem.* **2016**, *291*, 21616–21629. [CrossRef]
80. Liu, L.; Feng, D.; Chen, G.; Chen, M.; Zheng, Q.; Song, P.; Ma, Q.; Zhu, C.; Wang, R.; Qi, W.; et al. Mitochondrial outer-membrane protein FUNDC1 mediates hypoxia-induced mitophagy in mammalian cells. *Nat. Cell Biol.* **2012**, *14*, 177–185. [CrossRef]
81. Wu, W.; Lin, C.; Wu, K.; Jiang, L.; Wang, X.; Li, W.; Zhuang, H.; Zhang, X.; Chen, H.; Li, S.; et al. FUNDC1 regulates mitochondrial dynamics at theER–mitochondrial contact site under hypoxic conditions. *EMBO J.* **2016**, *35*, 1368–1384. [CrossRef] [PubMed]
82. Chen, G.; Han, Z.; Feng, D.; Chen, Y.; Chen, L.; Wu, H.; Huang, L.; Zhou, C.; Cai, X.; Fu, C.; et al. A Regulatory Signaling Loop Comprising the PGAM5 Phosphatase and CK2 Controls Receptor-Mediated Mitophagy. *Mol. Cell* **2014**, *54*, 362–377. [CrossRef] [PubMed]
83. Schapira, A.H.V.; Cooper, J.M.; Dexter, D.; Clark, J.B.; Jenner, P.; Marsden, C.D. Mitochondrial Complex I Deficiency in Parkinson's Disease. *J. Neurochem.* **1990**, *54*, 823–827. [CrossRef] [PubMed]
84. Schapira, A.H.V.; Holt, I.J.; Sweeney, M.; Harding, A.E.; Jenner, P.; Marsden, C.D. Mitochondrial DNA analysis in Parkinson's disease. *Mov. Disord.* **1990**, *5*, 294–297. [CrossRef]
85. Janetzky, B.; Hauck, S.; Youdim, M.B.H.; Riederer, P.; Jellinger, K.; Pantucek, F.; Zöchling, R.; Boissl, K.W.; Reichmann, H. Unaltered aconitase activity, but decreased complex I activity in substantia nigra pars compacta of patients with Parkinson's disease. *Neurosci. Lett.* **1994**, *169*, 126–128. [CrossRef]
86. Parker, W.D., Jr.; Parks, J.K.; Swerdlow, R.H. Complex I Deficiency in Parkinson's Disease Frontal Cortex. *Brain Res.* **2008**, *1189*, 215–218. [CrossRef]
87. Jr, W.D.P.; Boyson, S.J.; Ba, J.K.P. Abnormalities of the electron transport chain in idiopathic parkinson's disease. *Ann. Neurol.* **1989**, *26*, 719–723. [CrossRef]
88. Benecke, R.; Strümper, P.; Weiss, H. Electron transfer complexes I and IV of platelets are abnormal in Parkinson's disease but normal in Parkinson-plus syndromes. *Brain* **1993**, *116*, 1451–1463. [CrossRef]
89. Bindoff, L.; Birch-Machin, M.; Cartlidge, N.; Parker, W.; Turnbull, D. Respiratory chain abnormalities in skeletal muscle from patients with Parkinson's disease. *J. Neurol. Sci.* **1991**, *104*, 203–208. [CrossRef]
90. Blin, O.; Desnuelle, C.; Rascol, O.; Borg, M.; Peyro Saint Paul, H.; Azulay, J.-P.; Billè, F.; Figarella, D.; Coulom, F.; Pellissier, J.F.; et al. Mitochondrial respiratory failure in skeletal muscle from patients with Parkinson's disease and multiple system atrophy. *J. Neurol. Sci.* **1994**, *125*, 95–101. [CrossRef]
91. Cardellach, F.; Martí, M.J.; Fernández-Solá, J.; Marín, C.; Hoek, J.B.; Tolosa, E.; Urbano-Márquez, A. Mitochondria1 respiratory chain activity in skeletal muscle from patients with Parkinson's disease. *Neurology* **1993**, *43*, 2258–2262. [CrossRef] [PubMed]
92. Beal, M.F. Mitochondria take center stage in aging and neurodegeneration. *Ann. Neurol.* **2005**, *58*, 495–505. [CrossRef] [PubMed]
93. Lin, M.T.; Beal, M.F. Mitochondrial dysfunction and oxidative stress in neurodegenerative diseases. *Nature* **2006**, *443*, 787–795. [CrossRef] [PubMed]
94. Krige, D.; Carroll, M.T.; Cooper, J.M.; Marsden, C.D.; Schapira, A.H.V.; (The Royal Kings Queens Parkinson Disease Research Group). Platelet mitochondria function in Parkinson's disease. *Ann. Neurol.* **1992**, *32*, 782–788. [CrossRef]
95. Keeney, P.M.; Xie, J.; Capaldi, R.A.; Bennett, J.P., Jr. Parkinson's Disease Brain Mitochondrial Complex I Has Oxidatively Damaged Subunits and Is Functionally Impaired and Misassembled. *J. Neurosci.* **2006**, *26*, 5256–5264. [CrossRef] [PubMed]
96. Surmeier, D.J.; Obeso, J.A.; Halliday, G.M.; Surmeier, D.J.; Obeso, J.A.; Halliday, G.M. Selective neuronal vulnerability in Parkinson disease. *Nat. Rev. Neurosci.* **2017**, *18*, 101–113. [CrossRef]
97. Langston, J.W.; Ballard, P.; Tetrud, J.W.; Irwin, I. Chronic Parkinsonism in humans due to a product of meperidine-analog synthesis. *Science* **1983**, *219*, 979–980. [CrossRef]
98. Heikkila, R.E.; Manzino, L.; Cabbat, F.S.; Duvoisin, R.C. Protection against the dopaminergic neurotoxicity of 1-methyl-4-phenyl-1,2,5,6-tetrahydropyridine by monoamine oxidase inhibitors. *Nature* **1984**, *311*, 467–469. [CrossRef]
99. Gainetdinov, R.; Fumagalli, F.; Jones, S.; Caron, M.G. Dopamine Transporter Is Required for In Vivo MPTP Neurotoxicity: Evidence from Mice Lacking the Transporter. *J. Neurochem.* **2002**, *69*, 1322–1325. [CrossRef]
100. Ramsay, R.R.; Kowal, A.T.; Johnson, M.K.; Salach, J.I.; Singer, T.P. The inhibition site of MPP+, the neurotoxic bioactivation product of 1-methyl-4-phenyl-1,2,3, 6-tetrahydropyridine is near the Q-binding site of NADH dehydrogenase. *Arch. Biochem. Biophys.* **1987**, *259*, 645–649. [CrossRef]
101. Panov, A.; Dikalov, S.; Shalbuyeva, N.; Taylor, G.; Sherer, T.; Greenamyre, J.T. Rotenone Model of Parkinson Disease. *J. Biol. Chem.* **2005**, *280*, 42026–42035. [CrossRef] [PubMed]
102. Javitch, J.A.; D'Amato, R.J.; Strittmatter, S.M.; Snyder, S.H. Parkinsonism-inducing neurotoxin, N-methyl-4-phenyl-1,2,3,6 -tetrahydropyridine: Uptake of the metabolite N-methyl-4-phenylpyridine by dopamine neurons explains selective toxicity. *Proc. Natl. Acad. Sci. USA* **1985**, *82*, 2173–2177. [CrossRef] [PubMed]

103. Nicklas, W.J.; Vyas, I.; Heikkila, R.E. Inhibition of NADH-linked oxidation in brain mitochondria by 1-methyl-4-phenyl-pyridine, a metabolite of the neurotoxin, 1-methyl-4-phenyl-1,2,5,6-tetrahydropyridine. *Life Sci.* **1985**, *36*, 2503–2508. [CrossRef]
104. Ramsay, R.R.; Salach, J.I.; Singer, T.P. Uptake of the neurotoxin 1-methyl-4-phenylpyridine (MPP+) by mitochondria and its relation to the inhibition of the mitochondrial oxidation of NAD+-linked substrates by MPP+. *Biochem. Biophys. Res. Commun.* **1986**, *134*, 743–748. [CrossRef]
105. Przedborski, S.; Jackson-Lewis, V.; Yokoyama, R.; Shibata, T.; Dawson, V.L.; Dawson, T.M. Role of neuronal nitric oxide in 1-methyl-4-phenyl-1,2,3,6-tetrahydropyridine (MPTP)-induced dopaminergic neurotoxicity. *Proc. Natl. Acad. Sci. USA* **1996**, *93*, 4565–4571. [CrossRef]
106. Perier, C.; Bove, J.; Wu, D.-C.; Dehay, B.; Choi, D.-K.; Jackson-Lewis, V.; Rathke-Hartlieb, S.; Bouillet, P.; Strasser, A.; Schulz, J.B.; et al. Two molecular pathways initiate mitochondria-dependent dopaminergic neurodegeneration in experimental Parkinson's disease. *Proc. Natl. Acad. Sci. USA* **2007**, *104*, 8161–8166. [CrossRef] [PubMed]
107. Perier, C.; Tieu, K.; Guégan, C.; Caspersen, C.; Jackson-Lewis, V.; Carelli, V.; Martinuzzi, A.; Hirano, M.; Przedborski, S.; Vila, M. Complex I deficiency primes Bax-dependent neuronal apoptosis through mitochondrial oxidative damage. *Proc. Natl. Acad. Sci. USA* **2005**, *102*, 19126–19131. [CrossRef] [PubMed]
108. Fabre, E.; Monserrat, J.; Herrero, A.; Barja, G.; Leret, M.L. Effect of MPTP on brain mitochondrial H2O2 and ATP production and on dopamine and DOPAC in the striatum. *J. Physiol. Biochem.* **1999**, *55*, 325–331.
109. Chan, P.; DeLanney, L.E.; Irwin, I.; Langston, J.W.; Di Monte, D. Rapid ATP Loss Caused by 1-Methyl-4-Phenyl-1,2,3,6-Tetrahydropyridine in Mouse Brain. *J. Neurochem.* **1991**, *57*, 348–351. [CrossRef]
110. Burté, F.; De Girolamo, L.A.; Hargreaves, A.J.; Billett, E.E. Alterations in the Mitochondrial Proteome of Neuroblastoma Cells in Response to Complex 1 Inhibition. *J. Proteome Res.* **2011**, *10*, 1974–1986. [CrossRef]
111. Fornai, F.; Schlüter, O.M.; Lenzi, P.; Gesi, M.; Ruffoli, R.; Ferrucci, M.; Lazzeri, G.; Busceti, C.L.; Pontarelli, F.; Battaglia, G.; et al. Parkinson-like syndrome induced by continuous MPTP infusion: Convergent roles of the ubiquitin-proteasome system and α-synuclein. *Proc. Natl. Acad. Sci. USA* **2005**, *102*, 3413–3418. [CrossRef] [PubMed]
112. Chung, E.; Choi, Y.; Park, J.; Nah, W.; Park, J.; Jung, Y.; Lee, J.; Lee, H.; Park, S.; Hwang, S.; et al. Intracellular delivery of Parkin rescues neurons from accumulation of damaged mitochondria and pathological α-synuclein. *Sci. Adv.* **2020**, *6*, eaba1193. [CrossRef] [PubMed]
113. Tieu, K.; Perier, C.; Caspersen, C.; Teismann, P.; Wu, D.C.; Yan, S.D.; Naini, A.; Vila, M.; Jackson-Lewis, V.; Ramasamy, R.; et al. D-beta-hydroxybutyrate rescues mitochondrial respiration and mitigates features of Parkinson disease. *J. Clin. Investig.* **2003**, *112*, 892–901. [CrossRef] [PubMed]
114. Li, Y.; Zheng, W.; Lu, Y.; Zheng, Y.; Pan, L.; Wu, X.; Yuan, Y.; Shen, Z.; Ma, S.; Zhang, X.; et al. BNIP3L/NIX-mediated mitophagy: Molecular mechanisms and implications for human disease. *Cell Death Dis.* **2021**, *13*, 14. [CrossRef] [PubMed]
115. Zilocchi, M.; Finzi, G.; Lualdi, M.; Sessa, F.; Fasano, M.; Alberio, T. Mitochondrial alterations in Parkinson's disease human samples and cellular models. *Neurochem. Int.* **2018**, *118*, 61–72. [CrossRef] [PubMed]
116. Kocaturk, N.M.; Gozuacik, D. Crosstalk Between Mammalian Autophagy and the Ubiquitin-Proteasome System. *Front. Cell Dev. Biol.* **2018**, *6*, 128. [CrossRef] [PubMed]
117. Betarbet, R.; Sherer, T.B.; MacKenzie, G.; Garcia-Osuna, M.; Panov, A.V.; Greenamyre, J.T. Chronic systemic pesticide exposure reproduces features of Parkinson's disease. *Nat. Neurosci.* **2000**, *3*, 1301–1306. [CrossRef]
118. Liou, H.H.; Tsai, M.C.; Chen, C.J.; Jeng, J.S.; Chang, Y.C.; Chen, S.Y.; Chen, R.C. Environmental risk factors and Parkinson's disease. *Neurology* **1997**, *48*, 1583–1588. [CrossRef]
119. Tanner, C.M.; Kamel, F.; Ross, G.W.; Hoppin, J.; Goldman, S.; Korell, M.; Marras, C.; Bhudhikanok, G.S.; Kasten, M.; Chade, A.R.; et al. Rotenone, Paraquat, and Parkinson's Disease. *Environ. Health Perspect.* **2011**, *119*, 866–872. [CrossRef]
120. Johnson, M.E.; Bobrovskaya, L. An update on the rotenone models of Parkinson's disease: Their ability to reproduce the features of clinical disease and model gene–environment interactions. *NeuroToxicology* **2015**, *46*, 101–116. [CrossRef]
121. Lapointe, N.; St-Hilaire, M.; Martinoli, M.; Blanchet, J.; Gould, P.; Rouillard, C.; Cicchetti, F. Rotenone induces non-specific central nervous system and systemic toxicity. *FASEB J.* **2004**, *18*, 717–719. [CrossRef] [PubMed]
122. Höglinger, G.; Féger, J.; Prigent, A.; Michel, P.P.; Parain, K.; Champy, P.; Ruberg, M.; Oertel, W.H.; Hirsch, E. Chronic systemic complex I inhibition induces a hypokinetic multisystem degeneration in rats. *J. Neurochem.* **2003**, *84*, 491–502. [CrossRef] [PubMed]
123. Stefanis, L. α-Synuclein in Parkinson's Disease. *Cold Spring Harb. Perspect. Med.* **2011**, *2*, a009399. [CrossRef]
124. Di Maio, R.; Barrett, P.J.; Hoffman, E.K.; Barrett, C.W.; Zharikov, A.; Borah, A.; Hu, X.; McCoy, J.; Chu, C.T.; Burton, E.A.; et al. α-Synuclein binds to TOM20 and inhibits mitochondrial protein import in Parkinson's disease. *Sci. Transl. Med.* **2016**, *8*, 342ra78. [CrossRef] [PubMed]
125. Elkon, H.; Don, J.; Melamed, E.; Ziv, I.; Shirvan, A.; Offen, D. Mutant and Wild-Type α-Synuclein Interact with Mitochondrial Cytochrome C Oxidase. *J. Mol. Neurosci.* **2002**, *18*, 229–238. [CrossRef]
126. Ludtmann, M.H.R.; Angelova, P.R.; Horrocks, M.H.; Choi, M.L.; Rodrigues, M.; Baev, A.Y.; Berezhnov, A.V.; Yao, Z.; Little, D.; Banushi, B.; et al. α-synuclein oligomers interact with ATP synthase and open the permeability transition pore in Parkinson's disease. *Nat. Commun.* **2018**, *9*, 2293. [CrossRef]

127. Robotta, M.; Gerding, H.R.; Vogel, A.; Hauser, K.; Schildknecht, S.; Karreman, C.; Leist, M.; Subramaniam, V.; Drescher, M. Alpha-Synuclein Binds to the Inner Membrane of Mitochondria in an α-Helical Conformation. *ChemBioChem* **2014**, *15*, 2499–2502. [CrossRef]
128. Ludtmann, M.H.; Angelova, P.R.; Ninkina, N.N.; Gandhi, S.; Buchman, V.L.; Abramov, A.Y. Monomeric Alpha-Synuclein Exerts a Physiological Role on Brain ATP Synthase. *J. Neurosci.* **2016**, *36*, 10510–10521. [CrossRef]
129. Reeve, A.; Meagher, M.; Lax, N.; Simcox, E.; Hepplewhite, P.; Jaros, E.; Turnbull, D. The Impact of Pathogenic Mitochondrial DNA Mutations on Substantia Nigra Neurons. *J. Neurosci.* **2013**, *33*, 10790–10801. [CrossRef]
130. Hsieh, C.-H.; Shaltouki, A.; Gonzalez, A.E.; da Cruz, A.B.; Burbulla, L.F.; Lawrence, E.S.; Schüle, B.; Krainc, D.; Palmer, T.D.; Wang, X. Functional Impairment in Miro Degradation and Mitophagy Is a Shared Feature in Familial and Sporadic Parkinson's Disease. *Cell Stem Cell* **2016**, *19*, 709–724. [CrossRef]
131. Singh, F.; Prescott, A.R.; Rosewell, P.; Ball, G.; Reith, A.D.; Ganley, I.G. Pharmacological rescue of impaired mitophagy in Parkinson's disease-related LRRK2 G2019S knock-in mice. *eLife* **2021**, *10*, e67604. [CrossRef] [PubMed]
132. Podlesniy, P.; Vilas, D.; Taylor, P.; Shaw, L.M.; Tolosa, E.; Trullas, R. Mitochondrial DNA in CSF distinguishes LRRK2 from idiopathic Parkinson's disease. *Neurobiol. Dis.* **2016**, *94*, 10–17. [CrossRef] [PubMed]
133. Gambardella, S.; Limanaqi, F.; Ferese, R.; Biagioni, F.; Campopiano, R.; Centonze, D.; Fornai, F. CCF-mtDNA as a potential link between the brain and immune system in neuro-immunological disorders. *Front. Immunol.* **2019**, *10*, 1064. [CrossRef] [PubMed]
134. Anichtchik, O.; Diekmann, H.; Fleming, A.; Roach, A.; Goldsmith, P.; Rubinsztein, D.C. Loss of PINK1 Function Affects Development and Results in Neurodegeneration in Zebrafish. *J. Neurosci.* **2008**, *28*, 8199–8207. [CrossRef] [PubMed]
135. Wu, S.; Lei, L.; Song, Y.; Liu, M.; Lu, S.; Lou, D.; Shi, Y.; Wang, Z.; He, D. Mutation of hop-1 and pink-1 attenuates vulnerability of neurotoxicity in C. elegans: The role of mitochondria-associated membrane proteins in Parkinsonism. *Exp. Neurol.* **2018**, *309*, 67–78. [CrossRef]
136. Clark, I.E.; Dodson, M.W.; Jiang, C.; Cao, J.H.; Huh, J.R.; Seol, J.H.; Yoo, S.J.; Hay, B.A.; Guo, M. Drosophila pink1 is required for mitochondrial function and interacts genetically with parkin. *Nature* **2006**, *441*, 1162–1166. [CrossRef]
137. von Coelln, R.; Thomas, B.; Savitt, J.M.; Lim, K.L.; Sasaki, M.; Hess, E.J.; Dawson, V.L.; Dawson, T.M. Loss of locus coeruleus neurons and reduced startle in parkin null mice. *Proc. Natl. Acad. Sci. USA* **2004**, *101*, 10744–10749. [CrossRef]
138. Itier, J.-M.; Ibáñez, P.; Mena, M.A.; Abbas, N.; Cohen-Salmon, C.; Bohme, G.A.; Laville, M.; Pratt, J.; Corti, O.; Pradier, L.; et al. Parkin gene inactivation alters behaviour and dopamine neurotransmission in the mouse. *Hum. Mol. Genet.* **2003**, *12*, 2277–2291. [CrossRef]
139. Goldberg, M.S.; Fleming, S.M.; Palacino, J.J.; Cepeda, C.; Lam, H.A.; Bhatnagar, A.; Meloni, E.G.; Wu, N.; Ackerson, L.C.; Klapstein, G.J.; et al. Parkin-deficient Mice Exhibit Nigrostriatal Deficits but Not Loss of Dopaminergic Neurons. *J. Biol. Chem.* **2003**, *278*, 43628–43635. [CrossRef]
140. Noda, S.; Sato, S.; Fukuda, T.; Tada, N.; Uchiyama, Y.; Tanaka, K.; Hattori, N. Loss of Parkin contributes to mitochondrial turnover and dopaminergic neuronal loss in aged mice. *Neurobiol. Dis.* **2019**, *136*, 104717. [CrossRef]
141. Pickrell, A.M.; Huang, C.-H.; Kennedy, S.R.; Ordureau, A.; Sideris, D.P.; Hoekstra, J.G.; Harper, J.W.; Youle, R.J. Endogenous Parkin Preserves Dopaminergic Substantia Nigral Neurons following Mitochondrial DNA Mutagenic Stress. *Neuron* **2015**, *87*, 371–381. [CrossRef] [PubMed]
142. Sliter, D.A.; Martinez, J.; Hao, L.; Chen, X.; Sun, N.; Fischer, T.D.; Burman, J.L.; Li, Y.; Zhang, Z.; Narendra, D.P.; et al. Parkin and PINK1 mitigate STING-induced inflammation. *Nature* **2018**, *561*, 258–262. [CrossRef] [PubMed]
143. Stauch, K.L.; Villeneuve, L.M.; Purnell, P.; Ottemann, B.M.; Emanuel, K.; Fox, H.S. Loss of Pink1 modulates synaptic mitochondrial bioenergetics in the rat striatum prior to motor symptoms: Concomitant complex I respiratory defects and increased complex II-mediated respiration. *Proteom. Clin. Appl.* **2016**, *10*, 1205–1217. [CrossRef] [PubMed]
144. Seet, R.C.S.; Lee, C.-Y.J.; Lim, E.C.H.; Tan, J.J.H.; Quek, A.M.L.; Chong, W.-L.; Looi, W.-F.; Huang, S.-H.; Wang, H.; Chan, Y.-H. Oxidative damage in Parkinson disease: Measurement using accurate biomarkers. *Free Radic. Biol. Med.* **2010**, *48*, 560–566. [CrossRef]
145. Chung, S.Y.; Kishinevsky, S.; Mazzulli, J.R.; Graziotto, J.; Mrejeru, A.; Mosharov, E.V.; Puspita, L.; Valiulahi, P.; Sulzer, D.; Milner, T.A.; et al. Parkin and PINK1 Patient iPSC-Derived Midbrain Dopamine Neurons Exhibit Mitochondrial Dysfunction and α-Synuclein Accumulation. *Stem Cell Rep.* **2016**, *7*, 664–677. [CrossRef]
146. Verstraeten, A.; Theuns, J.; Van Broeckhoven, C. Progress in unraveling the genetic etiology of Parkinson disease in a genomic era. *Trends Genet.* **2015**, *31*, 140–149. [CrossRef]
147. Corti, O.; Lesage, S.; Brice, A. What Genetics Tells us About the Causes and Mechanisms of Parkinson's Disease. *Physiol. Rev.* **2011**, *91*, 1161–1218. [CrossRef]
148. Grünewald, A.; Kumar, K.R.; Sue, C.M. New insights into the complex role of mitochondria in Parkinson's disease. *Prog. Neurobiol.* **2018**, *177*, 73–93. [CrossRef]
149. Hu, Q.; Wang, G. Mitochondrial dysfunction in Parkinson's disease. *Transl. Neurodegener.* **2016**, *5*, 1–8. [CrossRef]
150. Krohn, L.; Öztürk, T.N.; Vanderperre, B.; Bencheikh, B.O.A.; Msc, J.A.R.; Laurent, S.B.; Spiegelman, D.; Postuma, R.B.; Arnulf, I.; Hu, M.T.M.; et al. Genetic, Structural, and Functional Evidence Link *TMEM175* to Synucleinopathies. *Ann. Neurol.* **2019**, *87*, 139–153. [CrossRef]

151. Bras, J.; Guerreiro, R.; Hardy, J. SnapShot: Genetics of Parkinson's Disease. *Cell* **2015**, *160*, 570–570.e1. [CrossRef] [PubMed]
152. Nakamura, K.; Nemani, V.M.; Azarbal, F.; Skibinski, G.; Levy, J.M.; Egami, K.; Munishkina, L.; Zhang, J.; Gardner, B.; Wakabayashi, J.; et al. Direct Membrane Association Drives Mitochondrial Fission by the Parkinson Disease-associated Protein α-Synuclein. *J. Biol. Chem.* **2011**, *286*, 20710–20726. [CrossRef]
153. Papkovskaia, T.D.; Chau, K.; Inesta-Vaquera, F.; Papkovsky, D.; Healy, D.G.; Nishio, K.; Staddon, J.; Duchen, M.; Hardy, J.; Schapira, A.; et al. G2019S leucine-rich repeat kinase 2 causes uncoupling protein-mediated mitochondrial depolarization. *Hum. Mol. Genet.* **2012**, *21*, 4201–4213. [CrossRef] [PubMed]
154. Ramonet, D.; Podhajska, A.; Stafa, K.; Sonnay, S.; Trancikova, A.K.; Tsika, E.; Pletnikova, O.; Troncoso, J.C.; Glauser, L.; Moore, D.J. PARK9-associated ATP13A2 localizes to intracellular acidic vesicles and regulates cation homeostasis and neuronal integrity. *Hum. Mol. Genet.* **2011**, *21*, 1725–1743. [CrossRef]
155. Wang, W.; Wang, X.; Fujioka, H.; Hoppel, C.; Whone, A.L.; Caldwell, M.A.; Cullen, P.J.; Liu, J.; Zhu, X. Parkinson's disease–associated mutant VPS35 causes mitochondrial dysfunction by recycling DLP1 complexes. *Nat. Med.* **2016**, *22*, 54–63. [CrossRef]
156. Apicco, D.J.; Shlevkov, E.; Nezich, C.L.; Tran, D.T.; Guilmette, E.; Nicholatos, J.W.; Bantle, C.M.; Chen, Y.; Glajch, K.E.; Abraham, N.A.; et al. The Parkinson's disease-associated gene ITPKB protects against α-synuclein aggregation by regulating ER-to-mitochondria calcium release. *Proc. Natl. Acad. Sci. USA* **2020**, *118*, e2006476118. [CrossRef] [PubMed]
157. Agarwal, D.; Sandor, C.; Volpato, V.; Caffrey, T.M.; Monzón-Sandoval, J.; Bowden, R.; Alegre-Abarrategui, J.; Wade-Martins, R.; Webber, C. A single-cell atlas of the human substantia nigra reveals cell-specific pathways associated with neurological disorders. *Nat. Commun.* **2020**, *11*, 4183. [CrossRef] [PubMed]
158. Um, J.-H.; Yun, A.J. Emerging role of mitophagy in human diseases and physiology. *BMB Rep.* **2017**, *50*, 299–307. [CrossRef]
159. Clark, E.H.; de la Torre, A.V.; Hoshikawa, T.; Briston, T. Targeting mitophagy in Parkinson's disease. *J. Biol. Chem.* **2021**, *296*, 100209. [CrossRef]
160. Dolman, N.J.; Chambers, K.M.; Mandavilli, B.; Batchelor, R.H.; Janes, M.S. Tools and techniques to measure mitophagy using fluorescence microscopy. *Autophagy* **2013**, *9*, 1653–1662. [CrossRef]
161. Dagda, R.K.; Rice, M. Protocols for Assessing Mitophagy in Neuronal Cell Lines and Primary Neurons. *Tech. Investig. Mitochondrial Funct. Neurons* **2017**, *123*, 249–277. [CrossRef]
162. Patergnani, S.; Bonora, M.; Bouhamida, E.; Danese, A.; Marchi, S.; Morciano, G.; Previati, M.; Pedriali, G.; Rimessi, A.; Anania, G.; et al. Methods to Monitor Mitophagy and Mitochondrial Quality: Implications in Cancer, Neurodegeneration, and Cardiovascular Diseases. *Methods Mol. Biol.* **2021**, *2310*, 113–159. [CrossRef] [PubMed]
163. Mauro-Lizcano, M.; Esteban-Martinez, L.; Seco, E.; Serrano-Puebla, A.; Garcia-Ledo, L.; Figueiredo-Pereira, C.; Vieira, H.L.; Boya, P. New method to assess mitophagy flux by flow cytometry. *Autophagy* **2015**, *11*, 833–843. [CrossRef] [PubMed]
164. Zhou, Z.D.; Xie, S.P.; Sathiyamoorthy, S.; Saw, W.T.; Sing, T.Y.; Ng, S.H.; Chua, H.P.H.; Tang, A.M.Y.; Shaffra, F.; Li, Z.; et al. F-box protein 7 mutations promote protein aggregation in mitochondria and inhibit mitophagy. *Hum. Mol. Genet.* **2015**, *24*, 6314–6330. [CrossRef] [PubMed]
165. Liu, Y.-T.; Sliter, D.A.; Shammas, M.K.; Huang, X.; Wang, C.; Calvelli, H.; Maric, D.S.; Narendra, D.P. Mt-Keima detects PINK1-PRKN mitophagy in vivo with greater sensitivity than mito-QC. *Autophagy* **2021**, *17*, 3753–3762. [CrossRef]
166. Suzuki, S.; Akamatsu, W.; Kisa, F.; Sone, T.; Ishikawa, K.-I.; Kuzumaki, N.; Katayama, H.; Miyawaki, A.; Hattori, N.; Okano, H. Efficient induction of dopaminergic neuron differentiation from induced pluripotent stem cells reveals impaired mitophagy in PARK2 neurons. *Biochem. Biophys. Res. Commun.* **2017**, *483*, 88–93. [CrossRef]
167. Sun, N.; Yun, J.; Liu, J.; Malide, D.; Liu, C.; Rovira, I.I.; Holmström, K.M.; Fergusson, M.M.; Yoo, Y.H.; Combs, C.A.; et al. Measuring In Vivo Mitophagy. *Mol. Cell* **2015**, *60*, 685–696. [CrossRef]
168. Kim, Y.Y.; Um, J.; Yoon, J.; Kim, H.; Lee, D.; Lee, Y.J.; Jee, H.J.; Jang, J.S.; Jang, Y.; Chung, J.; et al. Assessment of mitophagy in mt-Keima *Drosophila* revealed an essential role of the PINK1-Parkin pathway in mitophagy induction in vivo. *FASEB J.* **2019**, *33*, 9742–9751. [CrossRef]
169. Katayama, H.; Kogure, T.; Mizushima, N.; Yoshimori, T.; Miyawaki, A. A Sensitive and Quantitative Technique for Detecting Autophagic Events Based on Lysosomal Delivery. *Chem. Biol.* **2011**, *18*, 1042–1052. [CrossRef]
170. Sun, N.; Malide, D.; Liu, J.; Rovira, I.I.; A Combs, C.; Finkel, T. A fluorescence-based imaging method to measure in vitro and in vivo mitophagy using mt-Keima. *Nat. Protoc.* **2017**, *12*, 1576–1587. [CrossRef]
171. Winsor, N.J.; Killackey, S.A.; Philpott, D.J.; Girardin, S.E. An optimized procedure for quantitative analysis of mitophagy with the mtKeima system using flow cytometry. *BioTechniques* **2020**, *69*, 249–256. [CrossRef] [PubMed]
172. Allen, G.F.G.; Toth, R.; James, J.; Ganley, I. Loss of iron triggers PINK1/Parkin-independent mitophagy. *EMBO Rep.* **2013**, *14*, 1127–1135. [CrossRef]
173. Katayama, H.; Hama, H.; Nagasawa, K.; Kurokawa, H.; Sugiyama, M.; Ando, R.; Funata, M.; Yoshida, N.; Homma, M.; Nishimura, T.; et al. Visualizing and Modulating Mitophagy for Therapeutic Studies of Neurodegeneration. *Cell* **2020**, *181*, 1176–1187.e16. [CrossRef] [PubMed]
174. Khmelinskii, A.; Meurer, M.; Ho, C.-T.; Besenbeck, B.; Fuller, J.; Lemberg, M.K.; Bukau, B.; Mogk, A.; Knop, M. Incomplete proteasomal degradation of green fluorescent proteins in the context of tandem fluorescent protein timers. *Mol. Biol. Cell* **2016**, *27*, 360–370. [CrossRef] [PubMed]

175. Liu, J.; Liu, W.; Li, R.; Yang, H. Mitophagy in Parkinson's Disease: From Pathogenesis to Treatment. *Cells* **2019**, *8*, 712. [CrossRef]
176. Moskal, N.; Riccio, V.; Bashkurov, M.; Taddese, R.; Datti, A.; Lewis, P.N.; McQuibban, G.A. ROCK inhibitors upregulate the neuroprotective Parkin-mediated mitophagy pathway. *Nat. Commun.* **2020**, *11*, 88. [CrossRef] [PubMed]
177. McWilliams, T.; Prescott, A.R.; Allen, G.F.; Tamjar, J.; Munson, M.J.; Thomson, C.; Muqit, M.M.; Ganley, I.G. mito-QC illuminates mitophagy and mitochondrial architecture in vivo. *J. Cell Biol.* **2016**, *214*, 333–345. [CrossRef]
178. Ganley, I.G.; Whitworth, A.J.; McWilliams, T.G. Comment on "mt-Keima detects PINK1-PRKN mitophagy in vivo with greater sensitivity than mito-QC". *Autophagy* **2021**, *17*, 4477–4479. [CrossRef]
179. McWilliams, T.G.; Prescott, A.R.; Montava-Garriga, L.; Ball, G.; Singh, F.; Barini, E.; Muqit, M.M.; Brooks, S.P.; Ganley, I.G. Basal Mitophagy Occurs Independently of PINK1 in Mouse Tissues of High Metabolic Demand. *Cell Metab.* **2018**, *27*, 439–449.e5. [CrossRef]
180. Ashrafi, G.; Schlehe, J.S.; LaVoie, M.J.; Schwarz, T.L. Mitophagy of damaged mitochondria occurs locally in distal neuronal axons and requires PINK1 and Parkin. *J. Cell Biol.* **2014**, *206*, 655–670. [CrossRef]
181. Lin, T.-H.; Bis-Brewer, D.M.; Sheehan, A.E.; Townsend, L.N.; Maddison, D.C.; Züchner, S.; Smith, G.A.; Freeman, M.R. TSG101 negatively regulates mitochondrial biogenesis in axons. *Proc. Natl. Acad. Sci. USA* **2021**, *118*, e2018770118. [CrossRef] [PubMed]
182. Evans, C.S.; Holzbaur, E.L. Degradation of engulfed mitochondria is rate-limiting in Optineurin-mediated mitophagy in neurons. *eLife* **2020**, *9*, e50260. [CrossRef] [PubMed]
183. Harbauer, A.B.; Hees, J.T.; Wanderoy, S.; Segura, I.; Gibbs, W.; Cheng, Y.; Ordonez, M.; Cai, Z.; Cartoni, R.; Ashrafi, G.; et al. Neuronal mitochondria transport Pink1 mRNA via synaptojanin 2 to support local mitophagy. *Neuron* **2022**, *110*, 1516–1531.e9. [CrossRef] [PubMed]
184. Melser, S.; Lavie, J.; Bénard, G. Mitochondrial degradation and energy metabolism. *Biochim. Biophys. Acta* **2015**, *1853*, 2812–2821. [CrossRef] [PubMed]
185. Marinković, M.; Šprung, M.; Novak, I. Dimerization of mitophagy receptor BNIP3L/NIX is essential for recruitment of autophagic machinery. *Autophagy* **2020**, *17*, 1232–1243. [CrossRef]
186. Jiang, W.-X.; Dong, X.; Jiang, J.; Yang, Y.-H.; Yang, J.; Lu, Y.-B.; Fang, S.-H.; Wei, E.-Q.; Tang, C.; Zhang, W.-P. Specific cell surface labeling of GPCRs using split GFP. *Sci. Rep.* **2016**, *6*, 20568. [CrossRef]
187. Koker, T.; Fernandez, A.; Pinaud, F. Characterization of Split Fluorescent Protein Variants and Quantitative Analyses of Their Self-Assembly Process. *Sci. Rep.* **2018**, *8*, 5344. [CrossRef]
188. Cevheroğlu, O.; Kumaş, G.; Hauser, M.; Becker, J.M.; Son, D. The yeast Ste2p G protein-coupled receptor dimerizes on the cell plasma membrane. *Biochim. Biophys. Acta (BBA)-Biomembr.* **2017**, *1859*, 698–711. [CrossRef]
189. Bader, G.; Enkler, L.; Araiso, Y.; Hemmerle, M.; Binko, K.; Baranowska, E.; De Craene, J.-O.; Ruer-Laventie, J.; Pieters, J.; Tribouillard-Tanvier, D.; et al. Assigning mitochondrial localization of dual localized proteins using a yeast Bi-Genomic Mitochondrial-Split-GFP. *eLife* **2020**, *9*, e56649. [CrossRef]
190. Ryan, T.; Bamm, V.V.; Stykel, M.G.; Coackley, C.L.; Humphries, K.M.; Jamieson-Williams, R.; Ambasudhan, R.; Mosser, D.D.; Lipton, S.A.; Harauz, G.; et al. Cardiolipin exposure on the outer mitochondrial membrane modulates α-synuclein. *Nat. Commun.* **2018**, *9*, 817. [CrossRef]
191. Morales, I.; Sanchez, A.; Puertas-Avendaño, R.; Rodriguez-Sabate, C.; Perez-Barreto, A.; Rodriguez, M. Neuroglial transmitophagy and Parkinson's disease. *Glia* **2020**, *68*, 2277–2299. [CrossRef] [PubMed]
192. Joshi, A.U.; Minhas, P.S.; Liddelow, S.A.; Haileselassie, B.; Andreasson, K.I.; Dorn, G.W., II; Mochly-Rosen, D. Fragmented mitochondria released from microglia trigger A1 astrocytic response and propagate inflammatory neurodegeneration. *Nat. Neurosci.* **2019**, *22*, 1635–1648. [CrossRef] [PubMed]

MDPI
St. Alban-Anlage 66
4052 Basel
Switzerland
Tel. +41 61 683 77 34
Fax +41 61 302 89 18
www.mdpi.com

Cells Editorial Office
E-mail: cells@mdpi.com
www.mdpi.com/journal/cells

www.ingramcontent.com/pod-product-compliance
Lightning Source LLC
LaVergne TN
LVHW070632170726
843515LV00004B/233
* 9 7 8 3 0 3 6 5 5 0 1 5 2 *